ALSO BY JANET MARINELLI

The Naturally Elegant Home

The Environmental Gardener

Going Native: Biodiversity in Our Own Backyards

YOUR NATURAL HOME

WITHDRAWN

JANET MARINELLI AND
PAUL BIERMAN-LYTLE
Foreword by William Browning
and Amory Lovins

YOUR NATURAL HOME

A COMPLETE SOURCEBOOK AND

DESIGN MANUAL FOR CREATING

A HEALTHY, BEAUTIFUL,

ENVIRONMENTALLY SENSITIVE HOUSE

Little, Brown and Company

Boston   New York   Toronto   London

Copyright © 1995 by Paul Bierman-Lytle and Janet Marinelli

All rights reserved. No part of this book may be reproduced in any form or by any electronic or mechanical means, including information storage and retrieval systems, except by a reviewer who may quote brief passages in a review.

First Edition

Floor plan p. 61 by Clodagh. All other drawings by Paul Bierman-Lytle.

The authors would like to thank the following photographers and organizations for their kind permission to reproduce the photographs used in this book:

P. 1, Lizzie Himmel; p. 2, Fox Fibre; p. 3, Robert Perron; p. 4 top, Auro; p. 4 bottom and p. 5, Robert Perron; p. 6, Trus Joist Macmillan; p. 7 left, Don Carson; p. 7 right, Hebel USA; p. 8 top, Daniel Aubry; p. 8 bottom, Bosch; p. 9, Robert Perron; p. 10, Forbo; p. 11, Robert Perron; p. 12 top, Melva Levick; p. 12 bottom, Lizzie Himmel; p. 13, Daniel Aubry; p. 15, Auro; p. 16, Jack Lenor Larsen; p. 17 top, Don Carson; p. 17 bottom and p. 19, Robert Perron; p. 20, Merida Meridian; p. 21, Whirlpool; pp. 22 and 23, Don Carson; p. 24, Robert Perron; p. 26, Don Carson; p. 27, Robert Perron, p. 28, Don Carson; p. 29, Perma-Chink Systems, Inc.; pp. 32, 34, 35, 42, 44, 45, 53, 54, and 55, Robert Perron; pp. 60, 62 top and bottom, and p. 63 top, Daniel Aubry; p. 63 bottom, Lizzie Himmel.

Library of Congress Cataloging-in-Publication Data

Marinelli, Janet.
    Your natural home : a complete sourcebook and design manual for
creating a healthy, beautiful, environmentally sensitive house /
Janet Marinelli and Paul Bierman-Lytle ; foreword by Robert Browning and Amory Lovins. — 1st ed.
        p.   cm.
    Includes index.
    ISBN 0-316-09302-5 (hc) — ISBN 0-316-09303-3 (pb)
    1. House construction.   2. Green products.   3. Environmental
health.   I. Bierman-Lytle, Paul.   II. Title.
TH4812.M2367   1995   94-46327
690'.837 — dc20

10   9   8   7   6   5   4   3   2   1

RRD-OH

Published simultaneously in Canada by Little, Brown & Company
(Canada) Limited

Printed in the United States of America

To Colette, Jim and Nancy Chuda and The Colette Chuda Environmental Fund

The Colette Chuda Environmental Fund was established in 1991 to research the connection between childhood cancers and other diseases and environmental pollution. Through research and education the fund aims to protect children's inalienable right to good health through a clean and safe environment.

Indoor pollution and other environmental hazards are the focus of the fund's first study, "Handle With Care: Children and Environmental Carcinogens," recently published by The Natural Resources Defense Council.

Through the collaborative efforts of Jim and Nancy, and Olivia Newton-John, whose home, designed by Jim, is shown on the cover of this book, the message that common home building and decorating products can be hazardous to human health is reaching the public. The growing awareness of such issues is forming the groundwork necessary to help save children's lives.

It is in the fondest memory of Colette, who lost her life to cancer at the age of five, that we dedicate this book.

For further information, contact The Colette Chuda Environmental Fund, 8409 Yucca Trail, Los Angeles, CA 90046; (213) 656-8715.

SEP – 1997

CONTENTS

FOREWORD

Winston Churchill said that "we shape our dwellings, and afterwards our dwellings shape our lives." In recent years we have found this aphorism to be truer than Churchill probably realized.

Through a series of case studies, we have documented that energy-efficient buildings and office design offer the possibility of significantly increased worker productivity. Through improvements in lighting, heating, and cooling, workers can be made more comfortable and productive. On an annual basis, energy costs in commercial buildings average $1.81 per square foot, while office workers' salaries, benefits, and equipment are around $200 per square foot. An increase of one percent in productivity can save a company more than the entire energy bill. Enhancing the energy efficiency and livability of buildings through better design is therefore a powerful way to save money and improve the quality of life. Efficient design practices are cost-effective just from their energy savings; the resulting productivity gains make them indispensable.

Over 30 percent of total U.S. energy usage, 60 percent of the country's electricity, 60 percent of its financial resources, and 26 percent of the contents of its landfills are linked to buildings. Moreover, 85 percent of the average American's time is spent indoors. In many buildings the air inside is more polluted than the air outside. Although the harm from unsustainable buildings may appear to be localized, it actually ripples outward, sometimes for thousands of miles. Homes in Southern California are framed with old-growth lumber from Washington and powered by burning coal strip-mined from Navajo sacred lands in Arizona. Ultimately, the costs of poor design are borne not solely by a building's owner and those who work and live there but by everyone alive, and even by generations yet unborn.

Clearly more than just productivity is an issue. Our health and well-being, the health of those who build our buildings, and the health of the ecosystems where we build and from which our building materials are derived are at stake.

Environmentally sensitive design can significantly improve the comfort, aesthetics, resource efficiency, and value of properties while reducing pollution and saving money.

This book is a valuable survey of how to achieve those benefits in your home. The issues addressed include energy and water efficiency, indoor air quality, sustainably sourced materials, biological waste treatment, and even constructed habitats.

These concepts are the result of a new view of ecology and design. This is an emerging practice in which design and real estate development are used as tools for restoring ecosystems and human communities, improving the comfort and performance of workplaces and homes, and strengthening the profitability of projects. Rather than just thinking about minimizing impacts, why not try to make the building process one of creative restoration for environments and the people who live within them?

Green buildings are not a frill, but a new way of building that has multiple benefits, including economic performance. The best examples have a beauty and magical quality that feeds the soul. Using resource efficiency, renewables, and sustainably sourced materials is not only important and good for Earth; it is a moral imperative. Why not start with your own home? Let us use tools like this book to build a greener world, one house at a time.

William Browning and Amory Lovins
Rocky Mountain Institute

YOUR NATURAL HOME

INTRODUCTION

I have been writing about environmental homes for seventeen years. Paul Bierman-Lytle has been designing and building them for just as long. Writing this book together was a natural collaboration.

Environmental home design has come a long way in the past seventeen years. Back in the late 1970s when Paul and I were starting out in our respective fields, the OPEC oil embargo had triggered a series of price shocks that sent home heating and cooling bills through the roof. While they sat in their cars for hours waiting for a tank of gas, homeowners muttered about the prospect of yet another teeth-chattering winter hovered around the wood-burning stove, and another sweltering summer with the air conditioner locked at eighty degrees. Engineers and architects began racking their brains to design houses that squeezed every last bit of energy out of each gallon of heating oil. They also began to take a serious look at houses powered by an environmentally benign, virtually unlimited source of energy — the sun. The houses that resulted *were* energy efficient, but they weren't about to win any beauty contests.

One 1970s-vintage environmental home I vividly recall was built by a young New Hampshire couple in accordance with the latest passive solar principles. For heating, it relied entirely on the orientation and building materials of the structure itself, not on boilers or other mechanical systems. For example, earth was piled up, or bermed, around the north, east, and west walls to insulate them from frigid blasts of northern New England winter air. The long south side of the house was made mostly of glass to let in the

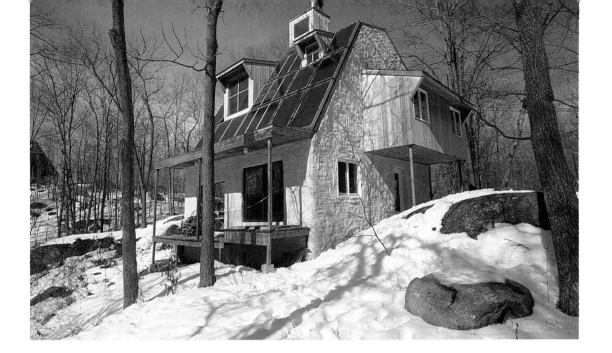

Environmental home design has come a long way. Solar houses of the 1970s, such as this Connecticut house, were energy efficient, but they weren't about to win any beauty contests.

warming sun. And just as attractive as its energy efficiency was the fact that this passive solar house did not need to rely on OPEC, oil companies, or local utilities. The house was so airtight that your ears popped when the front door slammed shut. But it wasn't very comfortable. The owners peeled off layers of clothing at noon when the sun sent indoor temperatures soaring, and then scurried for the down parkas at night when temperatures took a forty-degree nosedive. And the place looked about as inviting as a military bunker.

High-tech "active solar" designs, especially for heating domestic hot water, were also popular during the seventies. Thousands of houses around the country sprouted solar collectors that looked like alien space probes that had crashed into the roof. To make matters worse, many of these hastily introduced systems were lemons. It took the solar industry ten years to work the bugs out.

In the 1980s, health concerns came to the fore. The drive to install storm windows, plug air leaks, and stuff walls and attics with insulation trapped all that nice warm air inside the house — and a complex chemical soup of indoor air pollutants along with it. The "sick building syndrome" became a household term as newspaper headlines lurched from one indoor air-quality crisis to another. Some of the first complaints came from residents suffering from headaches, dizziness, irritation of the eyes and respiratory tract, and other troublesome symptoms. Formaldehyde levels in many

homes proved to be unusually high, and investigators found that the culprit was insulation made of urea-formaldehyde foam.

Subsequent research led to the discovery that formaldehyde, found in scores of the most ubiquitous home-building and decorating products, from plywood to fabrics to paints (and a suspected cancer causer), is only one of scores of so-called volatile organic compounds, mostly petrochemical-derived substances that readily volatilize, or become a breathable gas, at room temperatures and contaminate the air inside our homes. A growing body of scientific evidence suggests that these VOCs, as well as asbestos, radon, combustion gases from fuels used for heating and cooking, and pesticides routinely used indoors to control roaches, termites, and other pests, make the air inside our homes a much greater health risk than the air outside, even in the most industrialized cities. In the eighties a few environmental architects began grappling with the indoor air pollution problem, outfitting their homes with state-of-the-art ventilation systems and alternatives to the hundreds of potentially toxic building and decorating products found in the modern American house.

In the late 1980s and early 1990s ecological house design evolved by leaps and bounds. A few years ago, when I was researching my previous book, *The Naturally Elegant Home: Environmental Style,* I crossed paths with scores of architects and builders creating dwellings that are not only ecologically sound

▲
In the 1980s, environmental architects began outfitting their houses with alternatives to the hundreds of potentially toxic building and decorating products in the modern home. Today, homeowners can buy paints made of citrus oils, earth pigments, and other natural ingredients.

▼
A healthy kitchen includes cabinets constructed of solid wood and formaldehyde-free wood panels instead of polluting particleboard. Countertops are made of durable natural materials such as granite, not plastic laminates.

The new environmental home celebrates nature. Energy-efficient windows artfully frame views and flood the house with natural daylight, reducing the need for artificial lighting while making the living spaces bright and airy.

but also quite beautiful. These houses conserve energy, but unlike the environmental homes of the seventies they are not constrained by energy concerns. Space-age technologies, from superwindows to superinsulation, make them very responsive to the local climate. At the same time, however, they have all the emotional warmth of homespun architectural traditions and natural materials — materials that don't threaten the health of the inhabitants. Manufacturing these materials causes a minimum of pollution. Ideally, they're also "sustainable" — a fancy way of saying that they can be recycled or at least last as long as it will take nature to replace them in the wild.

The new environmental home is inspired by the natural character of the region. Stone, woods, and other materials native to the area reinforce its connection to the natural world. American beech, for example, a blond-colored wood suited to formal interiors, enhances the sense of living in the eastern deciduous forest, where this tree with smooth, silvery bark often attains impressive girth. It's also an excellent substitute for mahogany, an overexploited rainforest wood.

The state-of-the-art environmental home is inspired as well by the site itself. Great care is taken to preserve as much of the natural vegetation on the building site as possible; the floor plan may even be pushed or pulled to save a tree or rocky outcrop or arroyo. The house is also designed for smooth integration of water conservation and recycling into our busy everyday lives. In the new "green" kitchens, for example, handsome built-in cabinets discreetly store glass, aluminum, plastic, and other recyclables. In state-of-the-art environmental bathrooms, low-flow showerheads and ultra-low-flush toilets conserve every possible drop of precious water, and graywater systems whisk relatively unsullied water from the sink and shower into the garden for irrigation.

But these new environmental homes do more than protect nature; they celebrate it. Windows artfully frame views, and sun-

spaces allow us to live close to nature even in the dead of winter.
The profusion of glass also floods the house with natural light, reducing the need for artificial lighting while making the living
spaces bright and airy.

In its short four-hundred-year history, the American house has
been shaped by many styles: Georgian, Greek Revival, various Victorian styles from Italianate to Queen Anne, Shingle, Arts and
Crafts, Colonial Revival, Post-Modern, and more. During the past
decade, a new home style, the Environmental style, has begun to
come into its own. The Environmental style is still evolving. Its
proponents and practitioners are not yet recognized by scholars.
But already it is turning architecture on its head.

Most architectural styles are essentially veneers of fashion
cloaking a very conventional view of building. By contrast, the Environmental style embodies a far more fundamental set of principles. The goal of the Environmental style is to enrich human life,
to repair the ruptured link between us and the rest of nature, to
stimulate new ways of thinking and building and living that enable
us to coexist with the other creatures on Earth and to leave it in
better shape than we found it. This new style is spawning technologies and materials that one day will be the building blocks of
all architectural styles.

In the compendium of environmental building, decorating, and
maintenance materials that begins on page 77, we explore the materials and technologies used to build the new environmental

◀
**As the ancient forests of
the Pacific Northwest
have shrunk, so has the
supply of giant, old-
growth trees that have
been the mainstay of the
building industry.**

▲
**I-joists and other
composite wood
products make use of
lower grade wood fibers
and lumber mill scraps
that might otherwise be
discarded. I-joists can
carry the same load as —
or more than — solid-
sawn floor joists and roof
rafters, using about 50
percent less wood and
fiber.**

Rammed earth houses are made of earth, glass, and wood, materials that are renewed constantly by nature. Aerated concrete, used in the house at right, is handsome, conserves resources, resists rot, and does not burn.

home. Virtually every professional organization in the industry is taking a hard look at the environmental impacts of the various materials used to build our homes and other structures. For example, in 1992 the American Institute of Architects began publishing the *Environmental Resource Guide,* a groundbreaking quarterly review of building materials from an environmental perspective. Around the same time, the National Association of Homebuilders built the Resource Conservation Research House at its suburban Maryland headquarters, a rather imposing name for this yellow house with white trim and bay windows. However, about 80 percent of the products used to construct the house contain recycled materials. Industrywide, new product standards are being developed to take account of environmental concerns, while new building codes are being devised.

So-called life-cycle studies of the materials we use to build and decorate our homes — which evaluate and compare the entire range of environmental impacts of products from manufacture through disposal — will take years, if not decades, to complete. However, the basic criteria for environmental building products are clear: they do not pollute the indoor air, or at least produce substantially less indoor pollution than their conventional counterparts. They conserve energy and water. Buying or using them does not add to the pressure on threatened species or ecosystems. Mining and manufacturing them does not result in excessive air and water pollution. And because they are recyclable or biodegradable, disposing of them when their useful lives are over will not create environmental problems in the future.

One of the questions I'm most frequently asked about *The Naturally Elegant Home* is, "Where on earth do you get the nontoxic paints and natural linoleums that are in the houses in your book, or the beautiful rainforest woods that are harvested with a minimum of ecological destruction?" The present volume is designed to answer this question. In fact, the number of manufacturers producing environmental products is proliferating, both in this country and abroad. The pages that follow provide the most comprehensive homeowner's guide to environmental building, remodeling, and decorating products ever published — from tropical and domestic woods to wallpapers and paints to kitchen cabinets and countertops to fabrics and bedding. Hundreds of specific products are not only listed but also evaluated. Many

◀
If you want to make your home more environmentally healthy, the bedroom is a good place to start. Switch to untreated, 100 percent cotton sheets that, unlike polyester/cotton blends, are formaldehyde-free. Look, too, for super-energy-efficient appliances. The dishwasher below is also specially designed to conserve water.

▶
A painted floor is a charming alternative to wall-to-wall carpeting. Synthetic carpets can be a major source of potentially health-threatening pollutants. Even cotton, wool, and other natural carpeting can act as a "sink" for pollutants emitted by other sources, as well as for dirt, dust, dander, and other allergens.

of these products have been tested and used for years by Paul Bierman-Lytle's architectural and construction firm, The Masters Corporation, one of the world's leading pioneers of environmental building and remodeling materials.

In *Your Natural Home* we not only evaluate products but also provide you with tips on how to use them. And because many have not yet made it to your local Home Depot, we've also included the phone numbers and addresses of suppliers.

The philosophy implicit in the choice of materials recommended in this book is that it's better to be safe than sorry, especially when it comes to the health of our families. After all, we spend about 85 percent of our time indoors. And those most exposed to indoor air pollutants are those who are most vulnerable to their effects — children, the chronically ill, and the elderly. In recent years, scientists have discovered a great deal about the health hazards of some indoor pollutants, and they're learning more about others every day. There's solid evidence, for example, that radon is a major cause of lung cancer. On the other hand, we know much too little about the long-term health effects of the volatile organic compounds. The levels of some individual VOCs found in the typical house may not pose a significant health threat. However, the air inside most homes contains scores of VOCs. The combined effects of these pollutants may pose serious health risks. Choosing materials that minimize our exposure to VOCs and other potentially harmful pollutants when we build, renovate, or decorate is, we believe, the prudent course of action.

Children are especially vulnerable to many indoor pollutants because they have a higher metabolic rate than adults do, which means that they require more oxygen and therefore breathe in two to three times as much air (and air pollutants) relative to body size. For this reason, we've included a special section on how to create a house that is not only environment-friendly but also kid-friendly. Designed to let parents rest easy, this section offers common-sense advice on how to ensure that your home, particularly the nursery and playground, are free of lead paint, toxic pressure-treated woods, and other hazards to a child's health.

Your Natural Home is also an essential sourcebook of decorating ideas that are as handsome as they are healthy — both to us and the world we inhabit. The kind of fussy design so popular in neo-Victorian and "English country house" interiors of recent years can be not only excessive and wasteful but also expensive and time-consuming to maintain. In architecture and decoration — as in nature — restraint is often the most refined form of elegance. The kind of restraint we have in mind is anything but boring. In the bathroom of the New York City apartment featured on

Most "linoleum" available today is actually vinyl, a source of potentially hazardous indoor air pollutants. The manufacture of polyvinyl chloride, the main component of vinyl flooring, is one of the most toxic industries around. Natural linoleum, made primarily from powdered cork, linseed oil, and a backing of jute or burlap, is more durable than vinyl and is greaseproof, waterproof, and cushy underfoot. It is available as sheets or tiles in many colors, patterns, and textures.

materials that do not contribute to indoor air pollution and require little care. From paints made of plant oils in exquisite earth colors to area rugs made of untreated wools and rustic plant fibers to all-natural bedding, the hundreds of stylish decorating products recommended in the "Compendium" lay the groundwork for a new interior design that celebrates natural materials and superb craftsmanship.

In the bathroom of this New York City apartment, you won't find the balloon shades and overstuffed chaises found in opulent bathrooms of the 1980s. But you will find a delightful waterfall trickling down the stone wall behind the tub.

Another question I'm often asked is, "Don't environmental houses cost a lot more than conventional ones?" One of the most common criticisms leveled against environmental architecture is that it is too expensive for anyone except the very rich. But it doesn't have to be that way. The cost of environmentally sensitive materials varies widely. A good number of the products recommended in this book — low-biocide paint and natural bedding, to name just a couple — cost no more than their polluting counterparts. Other products — super-energy-efficient windows, for example — cost more initially but pay for themselves surprisingly quickly in lower fuel bills. Because environmental houses are designed for the local climate, they cost less not only to run but also to maintain. And they can be economical to build as well.

In Part One, "What's Wrong with Our Houses," you'll find a handsomely illustrated, easy-to-understand primer that points out where our homes' major environmental and health problems lie. In Part Two, "Building, Remodeling, and Decorating with Environmental Materials," we explain which conventional products pose the greatest health risks, how to set priorities when you build or remodel, and how to trim costs and make compromises judiciously so that you can have a healthy, ecologically sensitive, and beautiful house whether you're on a bare-bones, moderate, or spare-no-expense budget.

Of course, a home is much more than the sum of its parts. To show you just how beautiful an environmental home can be, we've included in-depth looks at how state-of-the-art materials and technologies fit together to create several handsome environmental residences — a new home, a renovation, a city apartment, and a weekend retreat.

We also explore the frontiers of ecological design. In the decades to come, we will face the depletion of some of the planet's most precious natural resources — from its oil to its bauxite to its bio-diversity — if our homes are not designed to give back to the Earth at least as much as they take. In "A Twenty-first-Century Home," we look at the spectacular living spaces that the new century holds in store.

PART ONE

WHAT'S WRONG

WITH OUR HOUSES

Our houses are our havens. Home has always been the quintessential symbol of a safe and protective environment — a place that shields us from a harsh climate and the dangers that lurk outdoors, that provides privacy, intimacy, and comfort, that nurtures health and family life.

In his book *Home: A Short History of an Idea* (1986), Witold Rybczynski describes how the idea of domestic comfort has changed over the centuries. In the seventeenth century, comfort meant privacy, which led to intimacy and, in turn, to the modern, middle-class notion of domesticity. In the eighteenth century, leisure and ease became important. In the nineteenth century, new domestic technologies led to a vast increase in levels of physical comfort. In the past hundred years, central heating, the flush toilet, running hot and cold water, electric lighting, and, finally, air conditioning have become middle-class amenities. During this century, domestic engineers have stressed efficiency and convenience. By the 1970s and eighties the American kitchen, for example, had become a repository for electrical appliances such as the microwave, designed for quick and easy processing of packaged convenience foods available at supermarkets.

Our homes have afforded us one of the highest living standards in the world. They're also at the very heart of complex ecological problems that threaten not only the health of the planet but ultimately our way of life.

Since the 1970s, we've begun trying to reconcile our need for comfort with concerns for the deteriorating environment. The infamous garbage barge, for example, which wandered up and down the East Coast for months trying to find a place to dump its

▶ Environmental kitchens are designed with materials that are as healthy as they are handsome: locally produced ceramic and quarry tiles, brick, slates, granites, solid woods, stainless steel, and terrazzo. Improved ventilation systems (*top*) get rid of harmful pollutants. The kitchen is also the center of energy and water conservation in the home.

odoriferous cargo of household trash, brought home the realization that we pay an environmental price for conveniences like packaging. Refrigerators and air conditioners have become necessities of modern life, but as we've learned in recent years, they also rely on a class of chemicals, the hydrochlorofluorocarbons (HCFCs), that have been linked to the destruction of protective ozone in the Earth's upper atmosphere. The massive oil spill from the *Exxon Valdez* in Alaska's Prince William Sound was another dramatic reminder of the environmental consequences of our voracious consumption of energy.

These and other lessons have encouraged us to make room in our kitchens for recycling. They've forced manufacturers to scramble to develop home appliances that don't depend on ozone-depleting chemicals. They've spurred us to make our homes more energy efficient. However, our houses still have a tremendous impact on the environment. And as we've learned during the past decade, they also have a tremendous impact on our own health.

Look around your house and you'll find sources of pollution almost everywhere. Although the health effects of these pollutants — and what you can do to reduce your exposure to them — will be discussed in detail in subsequent chapters, it's instructive to point out the major problems here.

One of the most serious household pollutants is radon. This radioactive gas, found naturally in rocks, soil, and groundwater, can seep into the house through cracks and drains, and can also be released into household air via tapwater. In tightly constructed homes in certain areas of the country, radon concentrations can be hundreds, even thousands, of times higher than background levels outside. Scientists consider radon a significant cause of lung cancer in this country.

The new petrochemical products we use on all the interior sur-

THE UNHEALTHY
HOUSE

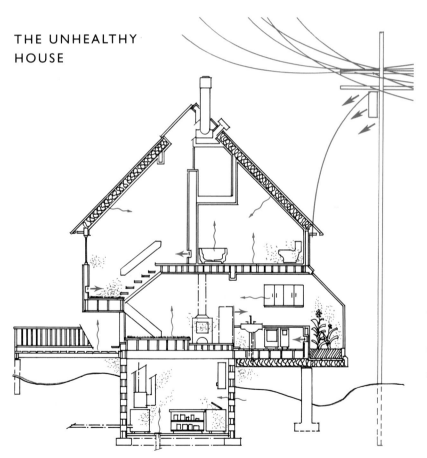

Scores of pollutants found throughout our houses can be hazardous to human health. Radon can seep into the house through cracks and drains and be released into the air via tapwater. The new petrochemical products we use on all the interior surfaces of our homes — from latex paints to synthetic carpeting — emit pollutants called volatile organic compounds, or VOCs. Nearby power lines and household electric appliances are sources of electromagnetic fields. What's more, our homes still suffer from an older generation of air pollutants, including carbon monoxide, molds, lead, and asbestos.

Whether for woodwork, floors, or furniture, sealers and finishes typically are petroleum based and can make the air inside your home unhealthy to breathe. On the spectacular woodwork in this study only natural finishes were used: these are made primarily of plant oils and beeswax, with which humans have lived safely for hundreds of years.

faces of our homes — from latex paints to synthetic carpeting — emit volatile organic compounds, or VOCs, which have been linked to a series of symptoms known as the sick building syndrome, including irritation of the eyes and respiratory system, headaches, even loss of coordination. Some VOCs, such as formaldehyde, are known or suspected cancer causers. There is little definitive information on the health effects of the levels of VOCs found in the average home. Nevertheless, the products that emit VOCs cover large areas of the house. We cover our walls and ceilings with latex paints, vinyl wallpapers, and wood paneling made with formaldehyde glues. We cover our floors with vinyl linoleums, vinyl tiles, and synthetic carpeting. We give our wood floors a polyurethane finish. From bureaus to bookshelves to kitchen cabinets, our furniture is likely to be constructed with plywood or particleboard, another big source of formaldehyde emissions. Up-

holstered furniture is padded with polyurethane foams and covered with fabrics
treated with chemical stain guards. VOCs
are also emitted from building materials
in out-of-the-way places — from the plywood or particleboard used for subflooring
and wall sheathing; from foam, fiberglass,
and cellulose insulation; from the myriad
adhesives used to construct the modern
house. And they're emitted as well from the
scores of products we use routinely for
cleaning and pest control. The newer or
more recently renovated your house is,
the higher the levels of VOCs are likely
to be.

As if this weren't enough, our homes still
suffer from an older generation of air pollutants. Gas stoves and boilers, kerosene
heaters, and wood-burning stoves and
fireplaces all release carbon monoxide,
nitrogen dioxide, and other harmful by-
products, especially if combustion is inefficient. Fungi, molds and mildews, pollen,
dust mites, and other allergens collect on wall-to-wall carpeting,
fabrics, and upholstered furniture.

Area rugs made from
jute, coir, sisal, and other
natural fibers are a rustic
alternative to wool and
cotton carpets. The
various plant fibers and
different weaves provide
a range of interesting
textures for your rooms.

And not only air pollutants pose health hazards in our homes.
The paint in older homes may be contaminated with lead. Lead
may leach out of old pipes into our drinking water. Our water supplies may also be tainted with nitrates from fertilizers, pesticides,
and industrial chemicals. Home wiring and all electric-powered
home appliances, from hair dryers to computers, emit electromagnetic fields. Scientists have yet to agree on the health hazards posed
by this electromagnetic radiation.

THE POLLUTING HOUSE

The typical house generates pollutants that threaten not only the
inhabitants but also the environment. What starts as a simple,
everyday act, such as lighting the fireplace, fertilizing the lawn, or

Our houses generate pollutants that threaten not only our health but also the environment. Our bathrooms and kitchens are the source of staggeringly large volumes of polluted water. The fertilizers and pesticides we use on our lawns and gardens can end up in underground water or nearby waterways. Our refrigerators and air conditioners rely on chemicals that destroy the protective ozone in the atmosphere. Our fireplaces and woodstoves, gas cooktops, and heating systems emit carbon dioxide and other global warming gases. And we still produce an astonishing amount of household trash.

THE POLLUTING HOUSE

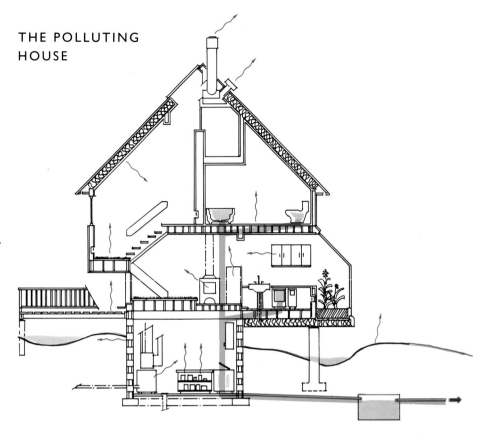

This Whirlpool refrigerator, which won a $30 million competition sponsored by the Environmental Protection Agency, does not use Freon, an ozone-depleting hydrochlorofluorocarbon, or HCFC. The award money is being used to offer rebates to buyers.

flushing the toilet, often ends up contributing to complicated, sometimes even global, ecological problems.

Wastewater is one of the most massive pollutant loads produced by our households. Polluted water from our sinks, dishwashers, showers, and toilets has to be treated somewhere. If you live in a rural area, it ends up in a septic tank on your property, where solids settle to the bottom. Excess wastewater then flows to an underground leach field, where soil organisms break down pollutants. Sometimes contaminated water reaches underground water supplies before it has been adequately broken down. And every few years, the septic tank needs to be pumped out. The septage is trucked to a "treatment" facility, often a series of primitive settling ponds or lagoons.

In large towns and cities, wastewater is piped to a sewage treatment plant, where solids settle out and bacteria begin breaking down pollutants. The solids, or sludge, are usually either incinerated or landfilled at enormous expense to the taxpayer. The "treated" water discharged from the treatment plant is still typically full of nitrogen and phosphorus that pollute receiving waterways.

The fertilizers and pesticides we use on our lawns and gardens can trickle down through the soil and into underground water, or run off our properties and end up in nearby waterways.

The hydrochlorofluorocarbons and chlorofluorocarbons released from our refrigerators and air conditioners, as well as those released during the manufacture of other home products such as foam insulations, end up in the atmosphere. There they can remain for a century or more, destroying ozone that protects us and the rest of the biosphere from damaging ultraviolet rays.

Our fireplaces, woodstoves, kerosene heaters, gas boilers, and gas cooktops and ovens produce not only health-threatening combustion pollutants but also carbon dioxide. Carbon dioxide is considered the most significant "greenhouse" gas and the leading contributor to the global warming that many scientists believe we are likely to witness in the coming decades.

According to studies commissioned by the Environmental Pro-

HOW MUCH GARBAGE DO WE GENERATE? (in pounds per person per day)	
1960	2.66
1970	3.27
1980	3.61
1988	4.00

(**Source:** U.S. Environmental Protection Agency)

Wastewater from our showers, sinks, dishwashers, and toilets is one of the most massive pollutant loads produced by our households. Low-flow showerheads and other efficient appliances and fixtures not only conserve fresh water but also cut down on the volume of wastewater that needs to be treated.

The massive walls of rammed earth houses such as this one are energy conserving and deaden noise, creating a hushed, soothing silence indoors. Even the floors are made from a resource-efficient soil-cement mixture that is scored to look like tile and grouted when dry.

COMPARISON OF GARBAGE GENERATION IN MAJOR WORLD CITIES
(in pounds per capita)

Los Angeles	6.4
Philadelphia	5.8
Chicago	5.0
New York	4.0
Tokyo	3.0
Paris	2.4
Toronto	2.4
Hamburg	1.9
Rome	1.5

(**Source:** National Solid Wastes Management Association)

tection Agency (EPA), the average U.S. resident produces four pounds of trash a day, or more than half a ton each year. Nationwide, says the EPA, our annual garbage output is more than 180 million tons. An overwhelming 73 percent of our household refuse ends up in overburdened landfills. Fourteen percent is incinerated. Only 13 percent is recycled.

The way we choose to design, build, and decorate our houses has enormous indirect pollution impacts as well. Some materials are extremely energy intensive or polluting to manufacture, causing environmental damage before they even get to our homes. A tremendous amount of energy is consumed in the manufacture of aluminum, for example. The manufacture of other materials, such as synthetic paints and plastics, generates a lot of toxic waste.

As homeowners and consumers, we can do a great deal to lessen this burden of pollution.

THE WASTEFUL HOUSE

It has become almost a cliché to note that our society is geared to consumption on a massive scale. Apparently the ramifications of this have yet to sink in. In our homes and our daily lives we continue to consume valuable, and in many cases irreplaceable, resources at a breathtaking clip.

The enormous waste of resources begins at the building site. One of the most valuable resources is land. The land a new house is built on is part of a complex, interconnected ecosystem. Yet often no consideration is given to preserving native vegetation at the site; trees and other plants are bulldozed at random. Precious topsoil is compacted and lost to erosion. The building site becomes a virtual landfill as materials are trimmed or discarded; little of this material is salvaged.

Salvaged woods, like the timbers used in this handsome house, have a distinct advantage: they are the highest-quality representatives of their species, having originated from old-growth forests many years ago, when trees twelve to fifteen feet or more in diameter were common. The heartwood of these old trees is unsurpassed.

Drought in the West, salt-water intrusion in the South, acid rain in the Northeast, and chemical contamination everywhere are making pure water an endangered resource. But we continue to consume prodigious amounts of this priceless natural resource. According to the EPA, the average American family of four uses 243 gallons of water indoors every day — 100 gallons for flushing the toilet, 80 for showering and bathing, 35 to do the laundry, 15 in the dishwasher, and another 13 in the sink. That's more than double the amount of water used by the typical European family. Every time we flush a conventional toilet, five to six gallons of water go down the tubes. The run-of-the-mill showerhead gushes five to eight gallons of water a minute. And these figures don't account

AN AMERICAN FAMILY'S DAILY WATER USE
(in gallons)

Activity	Water used
Toilet flushing	100
Showering and bathing	80
Laundry	35
Dishwashing	15
Bathroom sink	8
Utility sink	5

(**Source:** U.S. Environmental Protection Agency)

In our homes we continue to consume valuable, and in many cases irreplaceable, resources at a breathtaking clip. Often no consideration is given to preserving native vegetation and wildlife habitat on the land on which our houses are constructed. The average American family consumes more than twice the amount of pure water — an increasingly endangered resource — that the typical European family does. Although our houses have become more fuel efficient since the 1970s, they still waste far too much energy — most of it derived from nonrenewable fuels like oil and gas.

for the water used outdoors; yet our yards, especially our water-guzzling lawns, can account for as much as 50 percent of the total water used by the average family.

Although our houses have become more fuel efficient since the 1970s, they still waste far too much energy — almost all of it derived from nonrenewable fuels like oil and gas. Once these so-called fossil fuels, which were created as organic matter decayed over millions of years, are used up (and according to some estimates we will be out of cheap, accessible oil in thirty-five years), it will take millions more to replace them. According to the World Resources Institute, a Washington, D.C.–based environmental think tank, total annual energy use in the United States exceeds 81 quads, or 13.8 billion barrels

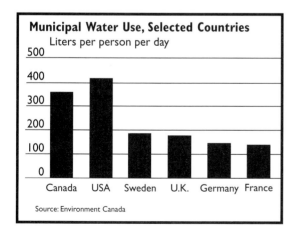

Municipal Water Use, Selected Countries

Liters per person per day

Source: Environment Canada

of oil. (A quad is 1,000,000,000,000,000, or one quadrillion, British thermal units of energy. A Btu, the standard measure of heat produced from various types of energy, is about equal to the energy of one burning match.) Our consumption of energy is still on the upswing — it has increased 9.7 percent since 1973, and 15.5 percent since a low in 1983. One eighth of our total energy use goes into heating and operating our houses.

U.S. ENERGY CONSUMPTION
(in quadrillion Btu)

Year	Btu
1973	74.282
1983	70.524
1990	81.453

(**Source:** Energy Information Administration, *Monthly Energy Review*, March 1991)

The waste of energy begins with the way we site and orient our homes. In cold climates, for example, we often fail to orient our houses toward the south to take maximum advantage of the natural heating benefits of the sun. We also fail to shelter our dwellings from cold prevailing winds. In hot climates, on the other hand, we fail to shade the house from the intense solar heat. Added to these fundamental flaws, inadequate insulation, gaps around doors and windows, and old, inefficient appliances all contribute to making our homes huge energy wasters.

◀

Energy experts calculate that more than half of the electricity we use on artificial lighting could be saved cost effectively if we converted to more efficient alternatives to the incandescent bulb, such as compact fluorescent and low-voltage halogen lights. This would cut in half the amount of fuel in power plants and dramatically reduce polluting air emissions. Our electric bills would be dramatically lower, too.

▶

The modern kitchen is our most direct link to the local landfill. We produce far more trash than the Japanese and Western Europeans do. Make room for recycling in your kitchen with attractive, built-in cabinets for temporary storage of cans, bottles, and other recyclable materials.

The sixteen pounds of solid waste generated by the average American family every day give the United States the dubious distinction of being one of the most wasteful countries on earth. Citizens of the United States generate twice as much garbage per person as individuals in Western Europe and Japan, according to the World Resources Institute. The amount of waste we create is growing every year. And every year there are fewer places to put it.

We can easily save water, energy, and other natural resources at home without crimping our lifestyles. The Europeans and Japanese use far less water and energy and create far less trash and yet still manage to live as well as we do. There are lots of good reasons to conserve resources. When you save energy in the home, for ex-

ample, you benefit the environment by reducing the pollution that results from the use of conventional energy sources. Many of our most pressing environmental problems — urban smog, acid rain, oil spills, and radioactive waste, to mention a few — are a direct result of energy production and use. You also save money on fuel bills — in many cases lots of it.

THE UNNATURAL HOUSE

The way our houses are designed affects not only our physical health but also our psychological well-being. Our homes have lost their once-intimate relationship with the land, which may contribute to the rootlessness and alienation that afflict modern life.

Lacking central heating, air conditioning, and other domestic technologies, the first European settlers on this continent had to get to know their climates intimately and make those climates work for them. New England "saltboxes" were built with a windowed two-story side to catch the sun and a low-sloping back roof turned to the prevailing winter winds. Chimneys and fireplaces formed a thick central core of masonry that retained heat and radiated it to surrounding rooms. Spanish settlers in the Desert Southwest, where the days are hot and dry and the nights are cool, constructed thick-walled adobe houses. The thick earthen walls soaked up solar heat, keeping the houses cool during the day and warm at night. During the past century, with the advent of central heating and cooling, however, Spanish villas have proliferated on Long Island, and saltboxes have been marketed in Texas, where they have no relation to the local climate or the architectural

▼
A house built of earth becomes part of the landscape rather than an imposition on it. Earth has been the most essential building material in the desert Southwest for 1,000 years — used first by the Pueblo Indians, later by the Spanish in their adobe structures, and today by builders who use pneumatic tampers to compact a soil-and-cement mixture between removable plywood frames.

▶
Log houses are romantic, but they are often made of woods in increasingly short supply. If you're building a log home, use a plentiful, fast-growing wood like aspen. Treat the logs with a low-toxicity wood preservative, and use a mortar that does not emit health-threatening pollutants.

heritage of the area. Inexpensive energy has meant we can keep our houses warm in winter and cool in summer, no matter how flimsy the walls are. Huge mechanical heating and cooling systems struggle to maintain comfort in houses ill suited to their locales.

Traditionally, houses were built using stone, woods, and other materials indigenous to the region. Whether fieldstone or terra-cotta, cedar, white pine, oak, or madrone, homeowners used materials from the local landscape, and their houses harmonized with nature. Today, plywood is trucked across the continent, and the synthetic, petroleum-based products ubiquitous in our living spaces have little relation at all to the natural world.

Our cocoon-like dwellings deny the sense of time as well as the sense of place. Modern houses force us to over-rely on artificial lighting, instead of encouraging us to celebrate the passage from dawn to dusk as the sun arcs across the sky.

At the very beginning of this century, some architectural critics were already bemoaning the boundaries that our homes erect between ourselves and the rest of nature. Gustav Stickley, publisher of *The Craftsman*, one of the most influential design magazines of the day, wrote in 1915, "Whatever connects a house with out of doors, whether vines or flowers, piazza or pergola, it is to be welcomed in the scheme of modern home-making. We need outdoor life in this country; we need it inherently because it is the normal thing for all people, and we need it specifically as a nation, because we are an overwrought people, too eager about everything except peace and contentment."

Since then, our need for nature and the "outdoor life" has become ever more urgent. Across a continent of breathtaking natural diversity and beauty we've created lookalike lawns and gardens that have little relation to the native plant communities of the regions in which we live. Few of us grow even a tiny fraction of our own food; for decades, the only link to the age-old rhythms of agriculture for the vast majority of Americans has been the produce section of the supermarket. In these fast-food times, even food preparation and mealtime are all but bereft of their former sensual pleasure.

Unlike conventional houses, the new environmental homes featured in the chapters that follow have evolved naturally from their site, climate, and region. They are a part of nature, not apart from it. They repair the ruptured link between nature and daily life.

PART TWO

BUILDING, REMODELING,

AND DECORATING WITH

ENVIRONMENTAL MATERIALS

The pages that follow feature several handsome environmental houses: a renovation, a new house, a weekend retreat, and a city apartment. Each one includes state-of-the-art environmental building and decorating products — products that are available right now and can make any house a healthy, ecologically sensitive, and spectacular place to live. (For more information on specific products, see the compendium of products that begins on page 77.)

Also featured is a house under construction in a canyon outside of Ketchum, Idaho. This dwelling, which marks the transition to a twenty-first-century architecture, offers a tantalizing glimpse of environmental living in the decades to come.

AN ENVIRONMENTAL RENOVATION

After living in her suburban Connecticut home for three years, Priscilla Toomey was ready to renovate. She had more than the run-of-the-mill remodeling in mind.

The Toomeys live in a two-story, four-bedroom, colonial-style home that was built in 1947. Priscilla was tired of an awkward layout that had the kitchen at the front of the house, next to a dark dining room and too far away from a family room at the rear. A working parent, she wanted an easy-care, state-of-the-art kitchen at the center of her home that would enable her to spend as much time as possible with her two young daughters, ages four and nine. The family room, the result of a remodeling that had enclosed a slate patio at the rear of the house and topped it with a flat roof that leaked, also required attention. But that's where the similarity to the typical home remodeling ends. One of the top priorities in the Toomey renovation was to rid the rooms of lead paint and other health-threatening toxins found in older houses, and to finish the interior with pollutant-free materials and products. The idea was to make the home not only more handsome but also healthier.

Most remodelings begin with drawn-out discussions about possible changes in design and layout. The Toomeys began with a diagnosis of the house's health problems. The home was tested for possible high levels of radon. The water was tested for lead and chemical pollutants. A comprehensive visual inspection of the structure was also performed to check for old lead paint, crumbling asbestos, and other hazards.

This comprehensive home health checkup turned up a number

of problems. The roof, for example, was made of old asbestos shingles. Walls and ceilings were covered with layer upon layer of old paint containing lead. The house was damp and dark, and so there was a great deal of mold and mildew. To make matters worse, the structure suffered from a lack of ventilation: there weren't enough windows, and there was no mechanical system to cleanse stale, polluted indoor air. The attached garage was unvented, making it easy for polluted car exhaust to get into the house. Outside, shrubbery hugged the siding, aggravating moisture problems inside the home. A sump pump worked nonstop because drainage problems were so bad. Thick, wall-to-wall shag carpeting created an ideal environment for dust mites, bacteria, and other allergens.

The new design for the Toomey home attended to these problems, making it healthier as well as more attractive and livable. Because the redesign called for dormers and other additions that necessitated breaking through the roof, the asbestos shingles were removed. (They would have been left in place if remodeling plans had not called for disturbing the roof because asbestos embedded in a hard matrix such as a roof shingle generally poses no health threat.) A licensed asbestos removal team was hired to do the work, and testing was done afterward to make sure that asbestos fibers had not contaminated the inside of the house. Where removal was necessary, old lead paint was stripped with the least toxic liquid products available; liquid strippers are preferable to sanding and other paint-removal methods because they generate a minimum of hazardous lead dust.

Because allergies run in the family, including allergies to dust and mold, great care was also taken to rid the Toomey house of these and other indoor pollutants. All the wall-to-wall carpets and underlayments were ripped out. An old hot-water heating system in the ceilings that leaked and promoted growth of mold and mildew was removed as well, along with existing leaky and water-wasting plumbing fixtures. A drainage system was installed around the perimeter of the house to direct rainwater away from the basement areas. Shrubbery crowded up against the house was dug up and replaced with a foot and a half of washed stone for moisture, insect, and rodent control.

The interior of the house was redesigned to take full advantage of nature's great cleansers and purifiers — fresh air and sunlight.

AN ENVIRONMENTAL RENOVATION

Renovation of the Toomey house began with a comprehensive health checkup. The house was dark and damp, and so there was a great deal of mold and mildew. To take full advantage of nature's great cleansers and purifiers—fresh air and sunlight—the remodeling added new bay windows and skylights. All wall-to-wall carpeting was ripped out, exposing oak floors. These were sanded and finished with natural oils and waxes instead of the typical polluting polyurethane.

Oriental rugs, steam cleaned to remove any chemical
treatments, add warmth to the rooms, and can be
taken outside for a good airing and shaking. To keep
down dust, window treatments are decidedly non-
fussy. In the upstairs and first-floor bedrooms, cozy
window seats were added where there were once no
windows at all. The new operable windows make the
most of natural ventilation in the warmer months and
flood the house with natural light all year long.

BEFORE

Family room

Living room

Dining room ←

Bedroom

Kitchen

Bedroom

Bedroom

AFTER

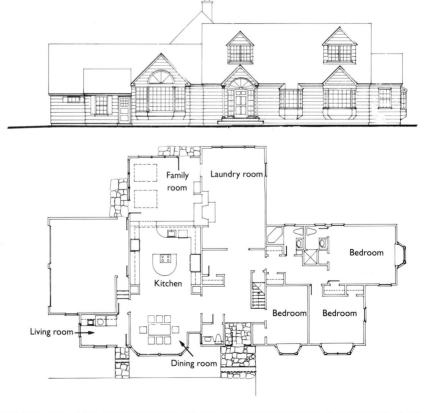

Family room

Laundry room

Bedroom

Kitchen

Bedroom

Bedroom

Living room →

Dining room

The remodeling added skylights, new bay windows, and dormers on the upper floors. On the first floor, for example, a twelve-foot-wide bay window opens up what has become the dining room to the daylight and the outdoors. A Palladian window plus side strip windows light up an elegant redesigned front entry. In the upstairs and first-floor bedrooms, cozy window seats were added where there were once no windows at all. The new operable windows make the most of natural ventilation in the warmer months and flood the house with natural light all year long.

The house was outfitted with a new central heating and cooling and air filtration system to keep mold, mildew, and other air pollutants and allergens to a minimum throughout the house year-round. The family's leaky old heating system was replaced with an efficient new gas-fired central heating and air-conditioning system. The new, state-of-the-art air filtration system consists of a HEPA (high-efficiency particulate absolute) screen to trap dust and other minute particles and an activated carbon medium to remove formaldehyde and other gaseous pollutants. A separate fan was installed in the garage to keep car fumes from polluting adjacent living spaces, and the door leading from the garage to the house was weatherstripped with airtight seals and gaskets. In all the newly remodeled areas, the old fiberglass insulation was removed, because its microscopic fibers can break down and disperse throughout the home, posing potential respiratory hazards. In its place, Air Krete, a noncombustible cementitious foam insulation made from a mineral derived from seawater, was installed.

These corrective measures made the house less hazardous. The next step was to help keep it healthy by using only building and decorating products that are petroleum-free or at least specially formulated to emit a minimum of formaldehyde and other volatile organic compounds. Removing the wall-to-wall carpeting exposed existing oak floors. These were sanded and finished with natural oils and waxes instead of the typical VOC-laden polyurethane. Oriental rugs, steam cleaned to remove any pesticides, mildew-cides, and chemical stain guards, add warmth to the rooms, and can be taken outside for a good airing and shaking. They also can be washed much more effectively than permanently affixed wall-to-wall carpet. A good-looking, moisture-proof slate floor was laid in the front vestibule without the usual toxic mastic or grouts. Wallpaper, which is also routinely treated with stain repellents and other chemicals, was used sparingly. Vinyl wallcoverings were avoided, and the wallpaper was applied with wheat paste instead of ready-mixed adhesives, which are highly polluting. German-made, petroleum-free paints made from plant oils and natural pigments were used on most walls as well as ceilings. Not only paints and floor finishes but also all adhesives and caulks used during the remodeling were selected to reduce indoor air pollution. The new environmental finishes are hard to tell from conventional products — they look great and are a lot healthier as well.

The most obvious transformation took place in the kitchen. The dining room and kitchen swapped locations, making the kitchen the focal point of the first floor. New cabinets, a table-sized island with a double sink, and ceramic tile floors make the kitchen even more inviting. Unlike the typical kitchen, which has cabinets and countertops made with particleboard, a major source of formaldehyde emissions, the Toomeys' are made of formaldehyde-free fiberboard with doors of solid cherry, an indigenous East Coast species, and finished with petroleum-free lacquer derived from plant oils. The new environmental kitchen was also outfitted with appliances that conserve energy and water.

In the family room, the leaking flat roof was replaced with a dramatic high-pitched ceiling pierced with skylights. These skylights, plus two pairs of sliding glass doors, make the formerly gloomy room bright and airy. Natural linoleum was installed on the family room floor. An extremely tough material, it can withstand a lot of wear and tear and, unlike vinyl linoleum, doesn't pollute the air with toxic, and potentially cancer-causing, VOC emissions.

As a finishing touch, incandescent light bulbs were replaced with energy-efficient compact fluorescents for general lighting throughout the Toomey home. High-performance and energy-conserving tungsten halogen fixtures were used for decorative and task lighting. Full-color-spectrum lights, which closely replicate natural daylight, were installed in the bathrooms and kitchen, where true color rendition is important.

From area rugs to window treatments, the interior furnishings add homeyness without being so elaborate as to become dust collectors. The striking new entrance, new bay windows, and high-pitched roof add character to the exterior of the house. The overall effect is both beautiful and healthy.

REMODELING ON A BUDGET
A Do-It-Yourselfer's Guide

One of the most effective ways to keep down remodeling costs is to do some of the work yourself. Here's a list of remodeling tasks any moderately handy homeowner can undertake, followed by a list of jobs best left to professionals. For detailed information on specific practices and products — from how to test radon levels in your house to the health advantages of natural linoleum over vinyl flooring — see the appropriate sections in the compendium of products beginning on page 77.

WHAT YOU CAN DO

Home Health Checkup

Test your home for radon, especially if you live in a high-risk area. Test for lead and chemical pollutants in your domestic water supply. Inspect the structure for lead paint, asbestos, mold and mildew, and carbon monoxide and other combustion gases escaping from your stove, oven, furnace, or boiler. Look for potential sources of formaldehyde. Be your house's own general physician, but consult specialists for complicated problems such as radon mitigation or asbestos removal if your initial testing turns up problems.

Floors

Remove wall-to-wall carpeting and vinyl tiles or sheet flooring, particularly if they have been installed recently (new synthetic carpeting and vinyl flooring emit the most VOCs). Replace them with natural linoleum or ceramic tiles. Refinish wood floors with natural, petroleum-free, or low-VOC synthetic wood finishes and waxes. Add warmth to your rooms with area rugs made of untreated wool, cotton, jute, sisal, coir, or other plant fibers.

Walls

Remove vinyl or chemically treated wallpaper, sources of polluting emissions, especially if they're new. Paint with natural or low-VOC paints. Build new walls using nonpolluting wallboard substitutes such as plaster over lath, cementitious fiberboards, or wallboard and low-VOC joint compound.

Wood Trim

Remove old paint with low-toxicity liquid paint strippers. Refinish with natural, petroleum-free, or low-VOC synthetic wood stains, oils and waxes, varnishes, or paints.

Cabinets and Built-Ins

Remove or seal particleboard, which emits urea-formaldehyde, the most hazardous form of formaldehyde. Replace with formaldehyde-free material and finish with natural, petroleum-free, or low-VOC synthetic oils or paints.

Lighting Fixtures

Replace incandescent light bulbs with energy-efficient compact fluorescent, tungsten halogen, and high-intensity discharge bulbs and fixtures and with healthy full-color-spectrum lamps where appropriate.

Plumbing Fixtures

Replace showerheads and faucets with water-conserving fixtures. Install a water filter if tests indicate that levels of a pollutant or pollutants in your tapwater are high.

The first step in any environmental renovation is to give your home a health checkup. Physically inspect your house for potential trouble spots: old lead paint and asbestos, for example, as well as mold and mildew, and building and decorating materials, such as particleboard and wall-to-wall carpeting, most likely to emit or be "sinks" for polluting volatile organic compounds. (See Appendix A, "A Guide to Home Pollutants," pages 245–249, for a list of materials and products most likely to emit formaldehyde and other VOCs.) Be sure to test for hidden health hazards: check radon levels in your home. Have your water tested for lead and chemical pollutants. If the home health examination turns up serious health risks, make them high priorities for corrective action during remodeling.

Which health hazards pose the greatest risks?

Radon should be at the top of your list of concerns, especially if you live in an area at high risk for the radioactive gas, considered a major cause of lung cancer in this country. (See the map on page 82.)

Lead, whether from peeling paint or drinking water, is the number-one health threat to children, because it can impair mental and physical development.

Deteriorating or damaged insulation or fireproofing material made of asbestos is another high priority, as inhalation of asbestos fibers has been shown to result in lung disease, from scarring of the lower lobes to cancers.

Less is known about the long-term health effects of some home pollutants, including the volatile organic compounds. However, studies give significant cause for concern, if not definitive proof. Formaldehyde is one of the most worrisome VOCs because it is in so many building and decorating products. It is a suspected cancer causer and has been linked to the development of chemical sensitivities in some individuals.

Appliances

Replace old appliances with energy-efficient, water-conserving, hydrochlorofluorocarbon-free (HCFC-free), and quiet models. This includes refrigerators, air conditioners, cooktops, ovens, freezers, microwaves, dishwashers, clothes washers, and dryers.

Noise

Outfit kitchen cabinets and doors with cork cushions or felt padding to deaden noise. Install felt weatherstripping on door jambs. Weatherstrip and insulate the laundry room, TV room, garage, and workshop. Insulate bedroom walls, ceilings, and floors. Install vibration cushions under appliances and mechanical equipment.

Recycling

Make room for an area in which to store used paper, metal, glass, and plastic containers for recycling and organic wastes for composting, preferably in the kitchen for optimum convenience.

Window Treatments and Other Decorating Fabrics

Replace vinyl shower curtains and chemically treated fabrics and upholstery with untreated natural fabrics from renewable sources, such as cottons, linens, silks, and rayons.

Bedding

Buy cotton sheets, pillowcases, pillows, and mattresses that haven't been treated with potentially health-threatening chemicals.

Computer and TV Screens

Install devices to shield you and your family from potentially health-threatening electromagnetic fields.

Maintenance Products

Instead of conventional, and generally highly toxic, cleaning products, use environmentally safe detergents, soaps, toilet cleaners, glass cleaners, polishes, waxes, and oils. Don't store these products close to foods or where they are easily accessible to children and pets. Avoid chemical lawn-care products, pesticides, herbicides, and other poisons in the home and the yard.

JOBS BEST LEFT TO PROFESSIONALS

Daylighting

As he or she redesigns your rooms, have your architect make the most of natural daylight, incorporating windows, skylights, dormers, bays, and French or sliding doors that complement the architectural style of your house. Invest in the most energy-efficient glazing you can afford.

Natural Ventilation

Make sure the renovation contributes to your home's ability to ventilate itself, with operable windows, screen doors, covered porches, and the like.

Heat-Recovery Ventilation

Ask your architect or heating contractor about installing a heat-recovery ventilator in your house. HRVs remove stale, polluted air and bring in fresh outside air. They also help control humidity and, in turn, mold and mildew problems when used in bathrooms, kitchens, and plant-filled sunrooms.

Ultra-Low-Flush Toilets

If you need a new toilet or your bathroom is being remodeled, have your plumber install a water-conserving ultra-low-flush model.

Energy Conservation

Consult experts about improving the energy performance of your home by checking for air leaks, adding insulation, or installing solar hot-water heaters and photovoltaic systems.

Electrical Wiring for EMF Reduction

While renovation is under way, have your electrician modify your home wiring, if necessary, to minimize your exposure to electromagnetic fields.

A NEW HOUSE

O n Buck Island, a tiny island off the Carolina coast, there is a home that exemplifies how the new environmental dwelling treads lightly on the land and nurtures the health of the people who live in it. The lush, thirty-five-acre, subtropical island can be reached only by boat. As the engines are cut and the boat glides silently to the dock, a ribbon of salt marsh comes into view. Beyond is a tangle of trees and coastal scrub — mostly palmettos and oaks draped with Spanish moss. Inserted between the trees on stilts that look like tree trunks is Jessica and Welles Murphey's house.

Trees and water are major design elements of this island abode. At the typical building site, trees are bulldozed like so many irksome obstructions, with little heed paid to the many roles they play in our land and lives. Trees are beautiful. Their leafy canopies cast dappled shade in summer. In many parts of the country, trees explode in a blaze of color in fall, and their bark adds substance and texture to the winter landscape. The sprightly yellow-green of trees leafing out brings cheer in early spring. Trees also provide essential habitat for birds and other creatures, absorb carbon dioxide and other pollutants, and create life-giving oxygen.

The design for the Murphey home began with a map of the two thousand or so trees on the island. The health of every tree was noted, and its trunk was measured. One of the first decisions made was that the structure would be built on stilts to minimize damage to the coastal forest. Raising the house above the ground would also aid in flood control, keep the living spaces above the dampness and insects, and encourage the sea breezes to wash the struc-

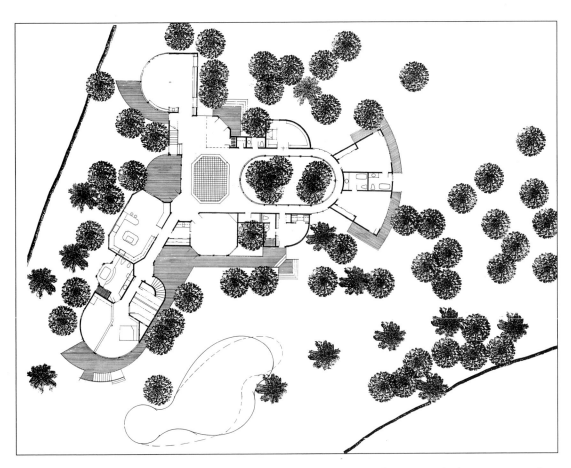

The design for the Murphey home began with a map of the trees on the island. In the gaps between the clusters of oaks and palmettos, a floor plan for the new house began to take shape.

ture with air, an important strategy for natural climate control in this hot and humid area.

On the map, random clusters of oaks and palmettos, known as hammocks, overlap a series of concentric circles, the first glimmer of a floor plan for the new house (see the above illustration). In gaps between the hammocks the living spaces began to take shape. The study, master bedroom, and exercise room extend between one corridor of trees, the living room and adjacent deck between another. The guest wing encircles still another grouping of palmettos, forming a native garden that can only be viewed from indoors. Decks flow around tree trunks and cascade down to the ground. Some trees grow out of the decks, and eaves were notched to let other specimens become a part of the home. Inside, a circular staircase spirals up to the "crow's nest," a romantic tower that soars above the tree canopy. A modern variation on the widow's walk of traditional seacoast homes, it is a dramatic spot from which to watch sunsets, approaching thunderstorms, and palmettos bending in the wind.

The house on Buck Island, a tiny island off the Carolina coast, is inserted between the trees on stilts that look like tree trunks. Inside, a circular staircase spirals up to the crow's nest, a tower that soars above the tree canopy. The home is a showplace of lesser-known tropical rainforest woods. Beyond the indoor garden (*opposite*) an existing clump of palmettos forms a native garden totally surrounded by the house's walls.

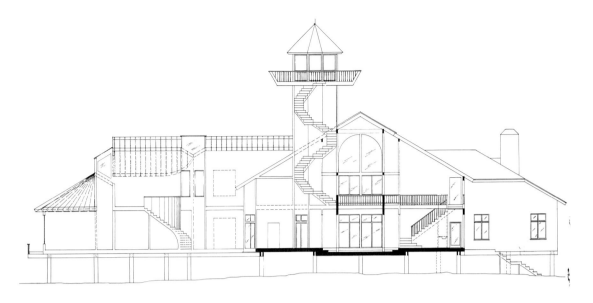

The Murphey home is a telling example of not only how trees can be preserved and celebrated at a building site but also how the trees brought in to build a structure should be carefully chosen. The 1x4 vertical tongue-and-groove siding is western red cedar, a wood well suited to the coastal Carolina climate. Highly rot resistant and insect repellant, cedar is long-lived and has beautiful grain figure. However, the magnificent old-growth specimens of this species, from which the highest-quality wood comes, have been severely depleted in the Pacific Northwest forests that are its native habitat. In 1984, when the Murphey house was constructed, every effort was made to buy the cedar from a reputable lumber company with a good environmental record. A decade later, we know that these slow-growing trees are being harvested faster than nature can replace them in the wild, and that further cutting puts pressure on disappearing old-growth forests. As our understanding of forest ecosystems has evolved, it has become apparent that the best course of action is to avoid using cedar altogether.

From the spectacular spiral staircase to the circular ceilings in the master bedroom and living room to the windows and doors throughout the home, the Murphey house is a showplace of tropical rainforest woods, particularly Brazilian cherry and African sipo mahogany. All decks are made of pau lope, a South American wood with natural preservatives that make it an excellent substitute for toxic pressure-treated lumber. Experts agree that using responsibly harvested timbers can add to the value of tropical

A modern variation on the widow's walk of traditional seacoast homes, the tower is a dramatic spot from which to watch sunsets, approaching thunderstorms, and palmettos bending in the wind.

ecosystems and therefore encourage husbandry of rainforest resources. All woods used on the interior of the Murphey house
were purchased from a company that claimed to be harvesting
them sustainably. At the time, however, there was no way to prove
that this was indeed the case. Today, architects and homeowners
can be absolutely certain that both tropical and domestic woods
have been harvested with a minimum of damage to forest ecosystems by buying them only from companies certified by Scientific
Certification Systems, the Rainforest Alliance, or other watchdog
group (see "Wood Products," pages 97–120).

Both interior and exterior woods were finished with low-
pollutant products made from linseed oil, citrus peel oil, and other
natural, nonpetroleum ingredients. The cedar siding was given
several coats of a natural plant oil and finished with two coats of
a natural teak-color stain that bring out the richness of the wood
and protect it from damage by the sun's ultraviolet rays. These natural finishes haven't been reapplied since the house was built. Indeed, they've stood up to the sun, wind, and salt spray of this
harsh coastal climate better than the petrochemical products that
are typically used in such settings. Indoor woods were coated with
liquid beeswax.

Both wood and water integrate the structure with nature. Water
is a prerequisite for life. It nourishes our bodies and sustains plants
and animals. It cools us. It offers opportunities for recreation. In
the words of Jacques Cousteau, the Earth, three quarters of which
is covered with water, is truly a water planet. Yet despite this seem-
ing abundance, freshwater use is increasingly outpacing supply
across the country, and it is difficult to find a source untainted by
sewage or industrial poisons. Freshwater is an especially precious
resource on a coastal island far from a public supply.

The Murphey house is designed to celebrate water and to use it
sparingly. From the crow's nest is a 360-degree view of the surrounding sea. The path to the front entry, paved with indigenous
oyster shells, skirts a picturesque lagoon, about as large as the
house itself. Two wells that tap into underground aquifers provide
the home with all the water necessary for drinking, cooking,
bathing, and swimming. They are also the home's main engines for
heating and cooling: a ground-source heat pump extracts heat
from the wells in winter to warm the rooms and cools them in
summer by transferring heat from the home to the underground

water. The ground-source heat pump also provides the Murpheys with all the domestic hot water they need. Recycled water from the ground-source system fills the lagoon.

Treated water from the toilets, sinks, and showers is recycled on the land. The Murphey house does not have a conventional septic tank and drainfield system, which can contaminate well water as wastewater percolates down through the sandy soil. Instead, used water is routed from the home to a septic tank, where some solids settle out. It is then treated in a microbial rock-bed filter along the lagoon, an area lined with an impermeable membrane and covered with rocks and gravel. Cleansed water flowing out of the rock-bed filter is used for irrigation. Planted with flowers, the rock bed looks more like a flowerbed.

HOME BUILDING ON A BUDGET

The three sample budgets that follow — a tight budget, a moderate budget, and a spare-no-expense budget — prove that it's possible to have a handsome, healthy, and energy-conserving house no matter how much money you make. There is some latitude in costs per square foot for each budget because incomes and costs vary across the country. For more information on specific recommendations, turn to the appropriate sections in the compendium of products beginning on page 77.

A TIGHT BUDGET

(\$30–\$80 per square foot)

If money is tight, you probably won't be able to hire architects or other consultants. You'll most likely be using blueprints from a plan book, magazine, or builder instead. But you can modify these plans to make the home more energy efficient and build and decorate it with materials that won't be hazardous to your health. Another way to save money is by doing a portion of the work yourself, particularly interior finishing. You can also be your own general contractor — which means that you'll be the one who hires subcontractors and monitors construction to make sure that it stays on time and on budget. But be forewarned: this is quite a time-consuming task and requires patience and organizational skills, as well as an understanding of building practices.

You won't be able to afford cutting-edge products and technologies if your budget is extremely tight. But even on a limited budget you can minimize indoor pollution and make your house a lot more energy efficient than the typical structure. Here's how:

■ Indoor Air-Quality Controls

- Reduce sources of indoor pollution by choosing products that emit a minimum of volatile organic pollutants for all the interior surfaces of your home. The following products are affordable even in this price rance: low-VOC caulks and sealants; flooring that does not contribute to indoor pollution, such as natural linoleum or ceramic tiles; low-VOC mastic and grouts; low-VOC joint compound; low-VOC paints; low-VOC wood finishes; low-VOC adhesives and glues; cabinetry and shelving constructed without particleboard and other materials that emit urea-formaldehyde.

- Furnish the house with area rugs made from wool, cotton, or other natural fibers instead of wall-to-wall carpeting.

- Buy untreated cotton sheets and pillowcases.

- Make sure your home is designed for natural ventilation, with plenty of windows that encourage cross ventilation in each room. Install bathroom, kitchen, and attic fans.

- Eliminate potential sources of combustion pollution. Don't build an attached garage from which car exhaust can easily infiltrate nearby living spaces. Consider alternatives to gas cooktops and ovens unless well-designed exhaust fans are installed along with them. Don't make wood-burning stoves or fireplaces your main form of heating.

- Install a HEPA (high-efficiency particulate absolute) air filtration unit in your central air ducting system, or at least in the bedrooms, and be sure to change the filters as directed.

- Don't use pesticides, herbicides, highly toxic cleaners, or other products that pollute your home with poisonous vapors.

Basic Energy Efficiency

- Locate your house to take full advantage of solar orientation and existing trees on site for shading and blocking wind (see "Heating and Cooling," pages 145–154).

- Make the most of passive solar heating and natural ventilation in your home's design. (This is also discussed in "Heating and Cooling.") Choose the most energy-efficient windows you can afford.

- Insulate your home well above recommended minimum levels.

- Buy energy-efficient appliances.

- Use compact fluorescent and other energy-efficient lighting where appropriate.

Recycling

- Organize your kitchen to make recycling paper, glass, metals, plastics, and other discarded materials convenient. A simple rolling cart with separate bins for the various materials is one inexpensive storage system for recyclables. Inexpensive stainless steel containers for temporary storage of food scraps destined for the compost pile are also available.

Environmental Landscaping

- Preserve as much of the native vegetation on the building site as possible.

- If you're building on a site in which the natural landscape is long gone, create a low-maintenance native garden by restoring forest, prairie, desert, coastal chaparral, or whatever plant community is indigenous to your area.

A MODERATE BUDGET

($80–$125 per square foot)

This price range will enable you to build in all the environmental features listed above. In addition, you'll be able to afford some or most of the following:

- More expensive natural products that do not pollute the air inside your home with the volatile chemicals in synthetics and are also petroleum-free. These include: natural, nonpetroleum paints; natural, nonpetroleum sealers and finishes; untreated wool and cotton carpets; untreated fabrics and upholstery; formaldehyde-free mattresses and box springs; low-pollutant insulation, such as Air Krete and Insulcot.
- Environmentally preferable structural components such as lightweight concrete, composite wood members, earth, and stone.
- Environmentally preferable exterior finishes, such as natural slate, wood fiber–cement composite slate or tile roofing, and natural paints or finishes for wood siding.
- Extremely energy-efficient windows, with energy-efficiency ratings of R-8 or more. (R-value is the measure of a material's ability to keep heat from flowing into or out of a room. Single-pane windows are rated R-1.)
- Appliances that are both energy-efficient and free of ozone-depleting chemicals.
- Ground-source heat pumps and solar hot-water heaters, which virtually eliminate the need for fossil fuels for heating and cooling.
- Hydronic heating, including hot-water radiant floors and baseboards, which are healthier and more energy efficient than forced air systems — and more comfortable.
- Heat-recovery ventilation plus combined HEPA and activated carbon air filtration for the highest-quality indoor air.
- Computer management systems for convenient and energy-efficient heating, cooling, and lighting controls, as well as home security.
- An integrated design for artificial lighting that includes compact fluorescent, tungsten halogen and high-intensity discharge lamps, and full-color-spectrum lighting.
- An area designed specifically to nurture a healthy body, which can include exercise equipment, whirlpool bath, steam shower, or even a sauna or lap pool.

A SPARE-NO-EXPENSE BUDGET

($125 per square foot and up)

When money is no object, you can build a home that includes cutting-edge systems, such as wastewater reclamation and solar electricity. With the help of consultants who are pioneers in their fields, you can also explore new technologies. In addition to those mentioned above, the following twenty-first-century components can be integrated into daily life:

- State-of-the-art structural materials, insulation, and glazing.
- Certified, salvaged, and naturally felled woods.
- The highest-quality natural paints and wood finishes.

- Ground-source heating, ice cooling, top-of-the-line air filtration and purification.
- A green kitchen complete with pantry, kitchen garden, and greenhouse for year-round production and storage of pesticide-free herbs, fruits, and vegetables, state-of-the-art HCFC-free, energy- and water-conserving, and totally quiet appliances, and automated home recycling center with chutes that channel recyclable materials to separate storage bins in the garage or outdoors.
- Indoor hydroponic gardens, aquaculture tanks for fish farming, and other "living machines" that provide food while purifying water from the showers, sinks, laundry, dishwasher, and toilets.
- Photovoltaic cells for solar electricity.
- Daylighting with a handsome, coordinated array of windows, French doors, atriums, and sunrooms.
- Furniture constructed with untreated fabrics and certified woods.

See also "A Twenty-first-Century Home," pages 67–76.

CHOOSING A SITE

Building an environmental home begins with careful site selection. You want a site with ample — and good quality — water and, if possible, trees. You can have an energy-efficient home if you choose a site with a microclimate that can work for you — for example, one with unobstructed access to solar energy. Because wilderness shrinks as uninhabited land is carved up into new home sites, it's important to preserve as much of the natural habitat as possible; better yet, avoid undeveloped areas and build in town. Be aware of any off-site hazards that can adversely affect indoor air quality. The following checklist can help you evaluate a potential building site. For more information, see the relevant sections in the compendium of products.

Habitat

- What vegetation exists on the site? It's a good idea to map the various plant communities. Inventory all trees with an eye toward designing a floor plan that disturbs as few as possible.
- Look for areas that are already disturbed or those where trees and other vegetation are sparse and can accommodate the footprint of a house. Locate your house as far as possible from wetlands, nests, and other areas of critical habitat for birds and other wildlife.

Water

- If the site is not serviced by a municipality or local water company, make sure that a water table is accessible for a well and that the water is potable.
- Visit the site when it's raining. You don't want to put a house where water collects and puddles up. By observing what happens to water on and around the

site, you can plan action to prevent water or moisture from entering the home, and collect it for other uses.

Microclimate

- Visit the site after a snowstorm. Does the snow melt more rapidly in some places than in others? If it does, this may be a sign of good solar orientation and you may want to put your house or greenhouse there.

- Observe the site when it's windy. Which direction does the wind come from? Do trees buffer the wind? Is there a cool side or warm side to the site where you might want to locate your house?

- Visit the site during the morning, at high noon, and in the late afternoon. Where is the sun? Does it penetrate the site or is it blocked by trees or other obstructions? By orienting your house toward solar south you can let the sun warm your rooms on winter days and save energy (see "Heating and Cooling," pages 145–154).

- Research the climate in the area. How much insulation is recommended? Knowing the average insolation levels (how much sunlight the area receives) will help you determine whether solar hot-water and electrical systems are feasible.

- Determine the category that best describes your climate: hot and humid, hot and arid, temperate and humid, temperate and dry, cold and humid, or cold and dry. This will help you design a structure and a mechanical system that is suited to the area (see "Heating and Cooling").

Potential Sources of Pollution

- Is the area at high risk for radon?

- Where is the nearest industrial facility? What kind is it?

- Are there any other potential sources of air or water pollution, such as nearby landfills, incinerators, or hazardous waste sites? Find out from local authorities if there has ever been a waste dump site nearby, or if there are plans to create one.

- Are there any major traffic arteries in the area?

- Are there any nearby facilities likely to be sprayed frequently with insecticides or herbicides — golf courses, nurseries, parks, farms, mosquito-breeding habitat, or corporate headquarters, for example?

- Check noise levels at the site. Is there an airport nearby? Is the site near a flight path? Are any major roadways in the vicinity? Manufacturing facilities, schools, and recreation areas can also generate noise pollution.

- How close are sources of powerful electromagnetic fields, such as transformers, transformer stations, high-voltage power lines, and railways and power plants?

A WEEKEND RETREAT

For several years, Bob and Barbara Raives, who spent weekdays in New York City, had been trekking up to a tiny house in rural Thetford Centre, Vermont, on weekends. The small cabin consisted of a combined living room, dining room, and kitchenette off a small mudroom entry and one bathroom. Above the bathroom and kitchenette was a bedroom loft reached by a ship's ladder. All of the electricity used in the cabin was generated by two solar photovoltaic panels mounted on the roof. The water heater, cooktop, and refrigerator were powered by propane fuel.

The Raiveses decided to build an extension onto the cabin to make room for children, nieces, nephews, and guests. Because the lot is small, the cabin had to be moved off its foundation to make room for the new addition. But the biggest challenge was to provide enough power for the enlarged living spaces, as the rural homesite is far from an electric utility grid. Typically, rural homes use wood for heating and cooking, propane for domestic hot-water heating and refrigeration, and a gas or diesel generator for electricity. The Raiveses wanted their home to be powered by the sun as much as possible.

In keeping with the architectural style of the cabin, the addition, which dwarfs the older structure, features wood siding, metal roofing, large trim boards, and double-hung windows. The effect is crisp, clean, and spare, like a Scandinavian farmhouse. The interior is also Scandinavian in its materials and lines. The walls and ceilings are poplar, left unfinished. Floors are maple and southern pine. All living spaces are flooded with natural daylight. The orientation is due south to take maximum advantage of passive solar heating.

Energy self-reliance is paramount in this remote mountaintop home. Even in a raw climate where sunshine levels are relatively low, solar electricity makes good sense. Photovoltaic panels are discreetly mounted flush with the metal roof. Two thirds of the fuel needed to heat the house is supplied by propane and wood. A highly efficient masonry heater (*left*) that forms the house's core releases heat from burning wood slowly, over several hours. It is a striking architectural feature that will last as long as the house.

An energy-efficient propane boiler and hydronic baseboard units are the house's primary heating system. In the bathroom, the hydronic heat comes from the sleek Runtel radiators behind the bidet. *Below:* The house is designed to make the most of solar heat. Twenty-one percent of the south-wall area consists of windows for optimum solar heat gain. This passive solar design provides a full one third of the energy needed to heat the home.

HOME HEAT LOSS SUMMARY

Building Component	Area (sq ft)	R-Value	Percent of Total
Roof	2,500.0	35.0	9.6
Floors			6.5
Doors	20.0	10.0	0.3
South walls	1,366.0	24.0	7.6
North walls	1,379.0	24.0	7.7
East walls	1,028.0	24.0	5.7
West walls	925.0	24.0	5.2
South glass	284.0	3.1	12.2
North glass	61.0	3.1	2.6
East glass	116.0	3.1	5.0
West glass	106.0	3.1	4.5
Infiltration	34,600.0 (cu ft)	0.4 ac/hr*	33.0

*air changes per hour

The Raives house is healthy through and through. All paints, adhesives, grouts, and caulking are either made with natural, non-petroleum ingredients or are specially formulated to minimize volatile organic pollutants. Woods were finished with pure tung oil or water-borne varnishes. No urea-formaldehyde-laden particle-board was used. Countertops are made of soapstone, instead of the usual plastic laminate over plywood, another source of urea-formaldehyde emissions indoors.

Energy self-reliance is paramount in the remote mountain home. The structure is highly insulated. To make room for extra insulation, it is constructed with 2x6 stud walls, instead of the standard 2x4s. The walls are filled with fiberglass batt insulation, for an R-value of 24; R-19 is typical for this climate zone. The roof is R-35, compared to the typical R-26, and attics are used as buffer spaces, not living spaces, increasing energy efficiency. What's more, the home is very tightly constructed, with few air leaks. All windows in the Raives house are low-emissivity (low-E), with argon gas between the panes; the low-E coating, an atoms-thin metallic film on the glass, reflects heat back into the house before it can escape, while the argon provides better insulation than air. The windows are rated R-3.1, compared to R-2 for standard, double-pane insu-

The Raives house is designed to capture solar heat. Twenty-one percent of the home's south-wall area (284 of a total of 1,366 square feet) consists of windows for good solar heat gain. At the same time, the house loses less than half the heat of the typical house of the same size. Only 4 percent of the cold north wall is glass, conserving energy. Walls are filled with six inches of fiberglass insulation for an R-value of 24 instead of the typical R-19. The roof is R-35, compared to the usual R-26. The house is extremely airtight, with little infiltration of cold outside air.

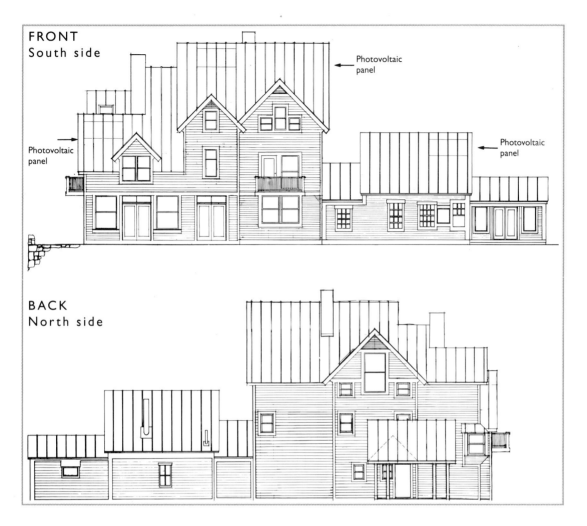

FRONT
South side

Photovoltaic panel

Photovoltaic panel

Photovoltaic panel

BACK
North side

lating glass. The overall result is that the Raives house loses 56 percent less heat than the typical home of the same size.

The structure is also designed to take advantage of solar heat. It is built on an east-west axis to face the sun. Twenty-one percent of the home's south-wall area consists of glass for optimum solar heat gain — five times more than the cold north wall, which has few windows to conserve energy. This passive solar design provides a full one third of the energy needed to heat the home.

To provide the remaining two thirds, the Raiveses opted for propane and wood. The primary heating system consists of a propane-fueled boiler and hydronic baseboard units throughout the house. The boiler has an integrated domestic hot-water unit that is more efficient than a stand-alone tank, and a respectable 83

WOOD AND PROPANE HEAT
A Comparison

Month	Cords of Wood	Gallons of Propane
January	1.8	280.4
February	1.4	221.8
March	1.0	159.4
April	0.4	66.6
May	0.1	13.1
June	0.0	0.0
July	0.0	0.0
August	0.0	0.0
September	0.0	3.5
October	0.3	42.1
November	0.9	137.2
December	1.6	254.9
Total Use	7.5	1,179.0
Annual Cost	$637.68	$1,414.75

Primary heating at the Raives house is provided by a propane-fueled boiler. Backup heat comes from a wood-burning masonry heater. Both the boiler and the masonry heater make efficient use of their respective fuels; more than 80 percent of the potential heat in the propane and the wood actually gets into the rooms. Propane costs more than wood for an equivalent amount of heating but is available at the flick of a switch. A masonry heater needs to be filled with wood at least twice a day.

percent overall efficiency rating. Secondary heat is provided by a masonry heater, which forms the house's core, radiating heat to the surrounding rooms. The masonry heater is a striking architectural feature that will last as long as the house. Highly efficient, it is designed to release the heat from the burned wood slowly, over several hours. Unlike a fireplace, it releases no unhealthy combustion fumes to the indoor air (see "Heating and Cooling," pages 145–154).

For electricity, the Raiveses decided to invest in solar. Even in this raw, northern New England climate where levels of sunshine are low compared to those elsewhere in the country, solar electricity makes good sense. Twenty-four modules — by no means an overwhelming array (see the photo on page 54) — are discreetly mounted flush with the metal roof. This provides a peak output of 1.35 kilowatts — enough to run all the home's appliances, including lighting, computers, television, refrigerator, well pump, and heating-equipment controls. A 2.5 kw converter changes the direct current produced by PV panels to standard alternating current. Twenty six-volt lead acid storage batteries provide 26.5 kwh of storage, which means that the system can run even on cloudy or rainy days. When the solar batteries are low, a backup propane generator kicks on; in case the owners are not at home, it has an automatic remote start. The cost of the system: $20,000, including controls, panels, batteries, generator, converter, and installation. However, with the cost of extending utility lines now running $40,000 per mile of line, the PV system is a bargain. And it is an even better deal when you figure that the cost is spread over the system's expected twenty-year life and that all the fuel needed to power the PV cells — sunshine — is free.

A few things would have been done differently had the budget allowed. The structure would have been even more highly insulated, with two-inch rigid board and cotton, rather than fiberglass, batt. All windows would have been Heat Mirror units, in which a plastic film with a low-E coating is stretched between the panes, boosting energy efficiency to an impressive R-8.

The propane-fired boiler, which uses a direct-vent duct system, eliminated the need for a costly chimney. However, experience has shown that it uses more electricity than a boiler with a chimney, adding to the load on the PV system. Even better, ground-source heat pumps, which extract warmth from the earth or a deep well, would have been used for heating instead of propane gas, so that the Raives house could have been heated entirely without fossil fuels.

A CITY APARTMENT

The new environmental home is designed to integrate nature and architecture, to restore human culture to its rightful place in the larger ecological community. Nowhere is this more necessary than in the city, where for blocks on end the only vestiges of wildness are often a struggling houseplant or street tree. It's not surprising, then, that a new interior design that emphasizes views and light while celebrating earthy materials and human craftsmanship is evolving in the midst of one of the planet's densest metropolises.

An apartment on Broadway and Sixty-eighth Street in Manhattan is the quintessential New York City space: on the top floor of a twenty-eight-story building, a short walk away from the hubbub of the theater district and the world-renowned cultural institutions at Lincoln Center. When the owners, a couple in their mid-forties, bought the apartment, it consisted only of a concrete slab in a building shell. This gave interior designer Clodagh virtually a blank canvas on which to work.

Because the owners are both gourmet cooks, the kitchen is at the center of the design. It is, literally, at the core of the space, where the mechanical systems of the two penthouses that were joined to form the apartment converged. Clodagh framed all the piping and ductwork, turning the kitchen into a striking pavilion flanked by Egyptianate columns. Adjacent to the kitchen is a huge greenhouse, which was raised in elevation to make the most of spectacular views of Broadway and the nearby Central Park Sheep Meadow. Banquettes invite the inhabitants to linger and enjoy the soaring skyline vistas. It's a wonderful place to have breakfast.

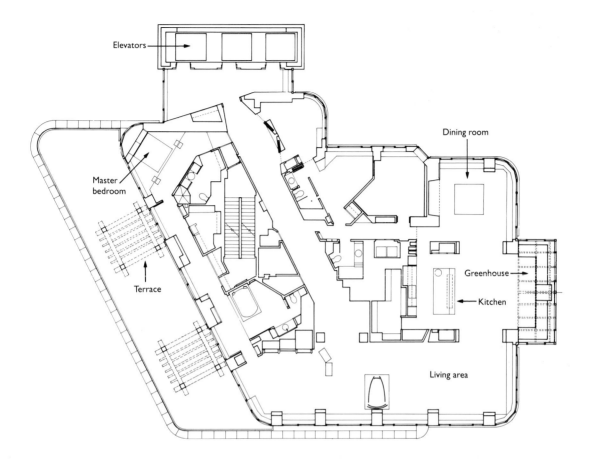

Elevators

Dining room

Master
bedroom

Greenhouse

Terrace

Kitchen

Living area

The indoor environment is designed to connect its urban dwellers to the earth. Walls and floors, the largest interior surfaces, are of paramount importance. The walls are plaster on lath; when mixed, the plaster was tinted a faded ocher. This color is kind to the eye and yet reflective, almost like sunshine: when daylight pours in from the many windows, the apartment is luminous, setting the bright and celebratory tone of the space. Warm oak flooring was used in many of the rooms. In the kitchen, a self-leveling concrete floor was ground so that it almost resembles a water silk pattern. The leather floor in the bedroom is aging like an old saddle. The flamed granite floor in one bathroom looks almost topographical, like stone eroded over time.

Every room is designed with what Clodagh calls a "ground." The focal point in the dining room (in fact, the only piece of furniture) is a heavy table made of purplish alcove stone from upstate New York. The impressive slab of stone rests on a rusted steel base. In most New York City apartments, doors are institutional

Adjacent to the kitchen is a huge greenhouse, which offers panoramic views of the nearby Central Park Sheep Meadow by day and Manhattan's twinkling skyline by night. Because the owners of the penthouse apartment are both gourmet cooks, the kitchen is at the center of the space, a striking pavilion flanked by Egyptianate columns that conceal piping and ductwork.

On summer evenings, the owners dine out on the twelve-hundred-square-foot terrace (*right*) where ninety different culinary herbs grow in a ziggurat of boxes. Indoors, designer Clodagh strives to connect the urban dwellers to the earth with what she calls a "ground" in every room. The focal point in the dining room (*above*) is a striking table made of purplish alcove stone from upstate New York. The impressive slab of stone rests on a rusted steel base.

and dreary; here, a massive oxidized copper door with handmade handle makes for a ceremonious entryway. In one bathroom a small waterfall trickles over rocks behind the tub. The modest-sized master bedroom is dominated by a bed that seems to float like a boat between windows and mirror on opposite walls. Ninety different culinary herbs grow in a ziggurat of boxes on the twelve-hundred-square-foot outdoor terrace.

Throughout the apartment, nothing is extraneous. There is no clutter. Doors, floors, dining table, and other functional furnishings are inseparable from the rooms. Much expression is achieved with few materials. The effect is both minimal and sumptuous.

Natural materials help ground the city dwellers. Stone, metal, plaster, and concrete in earthy colors and textures predominate. Very hard materials are used to surprisingly soft effect. These materials are inert, and therefore do not contribute to indoor air pollution. They don't require much maintenance, either. This not only makes life easier for the owners but also is easier on the environment, as the materials do not require constant applications of potentially toxic cleaners, polishes, waxes, and finishes. The beating, grinding, and other treatments to which the materials were subjected add patina. And because they already have a salvaged look, it doesn't matter if they get dinged up.

To eliminate dust and clutter, as much hidden storage as possible is built in. A line in the wall marking a drawer, a door handle, is all you see.

In this and other Clodagh interiors, low-maintenance fabrics predominate as well — leather, kilims and heavy raw silks, linens, and cottons. Soft furniture, used sparingly, is made exclusively with natural materials: goose down, cotton batting or kapok (a tropical plant fiber), horsehair, and natural jute webbing, instead of the usual synthetic foams and chemically treated fabrics.

The result is a living space that nurtures the body and soothes the soul, a natural house in the sky that is exquisitely crafted but not fussy — a perfect marriage of earthy materials and city sophistication.

TEN STEPS TOWARD AN ENVIRONMENT-FRIENDLY APARTMENT

Life in big cities has environmental advantages and disadvantages. On the minus side is the fact that urban dwellers usually don't have a big backyard or a natural area nearby. Noise — from screaming car alarms to loud neighbors — can drive you to distraction. On the other hand, life isn't driven by the automobile; you can walk to the store and in many cases walk to work. And multi-family buildings disturb much less land than the equivalent number of single-family dwellings.

Because you likely spend less time in your yard and more time indoors if you live in the city, it's important to have a home that is free of health-threatening pollutants. And because the way our cities traditionally have been organized and built alienates us from the larger fabric of life, it helps to design your surroundings in ways that remind you that humans are inextricably entangled with the rest of nature, products of natural evolution like every other species on the planet.

So how do you go about greening your city apartment?

1. City renovations, like environmental remodelings everywhere, should begin with a thorough home health checkup. Older buildings are probably full of old lead paint. Lead may be leaching out of old pipes and into your drinking water. Be sure to check for these and other potential health hazards, such as crumbling asbestos in old fireproofing and insulation. Check radon levels in your apartment if you live on the ground or first floor. Look for sources of volatile organic pollutants, including formaldehyde from particleboard and plywood in cabinets and other furniture and styrene butadiene in synthetic carpeting. See "An Environmental Renovation," pages 32–41; "A Guide to Home Pollutants," pages 245–249; and the relevant sections of the compendium of products for details on these and other home pollutants and what to do about them.

2. Insulate yourself as much as possible from noise pollution — from sirens, car alarms, and the din of traffic outdoors; from blasting stereos and wailing babies next door; and from noisy old refrigerators, dishwashers, and other appliances in your own apartment. Stuff your walls, ceilings, and floors full of insulation, particularly the bedrooms so you can sleep in peace. Weatherstrip your doors and windows. Install vibration cushions under all offending appliances.

3. As you renovate or remodel, take advantage of every opportunity to flood your living spaces with natural daylight. Work bay windows, skylights, greenhouses or other sunspaces, glass doors, and/or clerestory windows (windows located high up, where the walls meet the ceiling) into your redesign.

4. Invest in energy- and water-conserving appliances and fixtures. For example, replace incandescent lights with compact fluorescents and low-voltage halogens. If you're remodeling the bathroom, install an ultra-low-flush toilet.

Make one of the sleek new water-miser dishwashers a priority when you renovate the kitchen.

5. Make sure your apartment is properly ventilated. At a minimum you should have properly sized exhaust fans in the bathrooms and kitchen. Ideally, your apartment should be outfitted with a heat-recovery ventilator, which transfers the heat from stale, outgoing air to fresh, incoming air.

6. If the outdoor air is frequently polluted, you should also have a state-of-the-art air filtration system that combines a HEPA (high-efficiency particulate absolute) filter to screen out particulates and an activated-carbon medium to remove gaseous fumes.

7. Because the kitchen is increasingly the center of daily life, ask your architect or interior designer to design a cutting-edge green kitchen, complete with formaldehyde-free cabinets designed to make recycling easy, a refrigerator that is super-energy-efficient and does not rely on ozone-depleting chemicals, a water- and energy-efficient dishwasher, an excellent ventilation system to exhaust combustion pollutants from ranges and ovens, natural linoleum or ceramic tile flooring, a handsome array of windows to provide ample daylight, and a greenhouse window or attached sunspace so that you can grow your own fresh, pesticide-free herbs and salads.

8. Remodel with building and decorating products — from paints to caulks to carpeting — that are either natural and petroleum-free or else specially formulated to emit a minimum of volatile organic compound emissions.

9. Reinforce the connection between culture and nature by using natural materials dramatically throughout your home, including stone, woods, and natural-fiber fabrics. Grow an indoor herb garden, or re-create a patch of tropical rainforest in your living room.

10. Keep your apartment healthy by using cleaning and maintenance products that don't pollute your indoor air.

For more information on all the products mentioned above, see the compendium of products, beginning on page 77.

A TWENTY-FIRST-CENTURY HOME

In a canyon outside of Ketchum, Idaho, a home is emerging that marks the transition to a twenty-first-century architecture. In the 1920s, LeCorbusier declared that a house "is a machine to live in." The house in the Sawtooth foothills turns this statement on its head: like a handful of other structures on the cutting edge of ecological design, it is a living machine, a dwelling suffused with wild habitats that double as human life-support systems. In environmental homes of the late 1980s and the 1990s, the boundary between nature and architecture is blurred: opaque walls are replaced with long expanses of glass to bring the outside in and let the inside out. In the twenty-first-century home, this boundary disappears entirely. Ecological communities once confined to the outdoors now live within the house's walls.

At the dawn of the new century, the most urgent ecological crisis is the threat to biodiversity, due primarily to habitat destruction. In the United States, about four hundred square miles of undeveloped land is transformed into suburb every year; as wilderness shrinks, backyard acreage increases. Botanists are concerned about the long-term survival of one fifth of the native plant species in this country, plants that are critical habitat for countless other creatures. When we concentrate only on preserving isolated tracts of wilderness, creating, in the words of Smithsonian scientist Dr. Walter Adey, "an apartheid of human systems and wild systems," both suffer. A sustainable environment for the human species relies on a diversity of plant and animal species, as the twenty-first-century home attests.

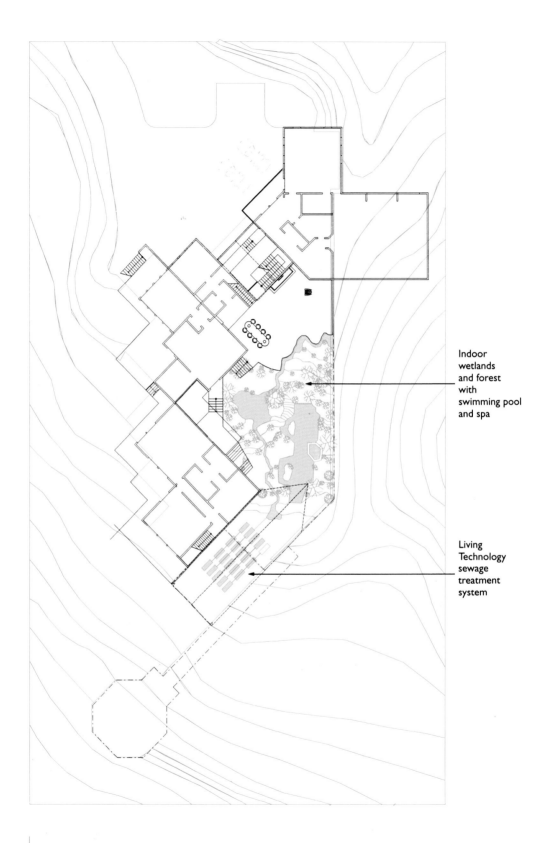

Indoor
wetlands
and forest
with
swimming pool
and spa

Living
Technology
sewage
treatment
system

The Idaho home is reached by a dirt road that snakes through the sagebrush scrub native to the high-desert environment. The house, called River's Edge, is nestled into a mountain, on a small plateau, elevation 5,700 feet. Just beyond, the land plunges steeply down to the Big Wood River.

Imagine for a moment that construction is completed: from a distance, the house is barely discernible from the surrounding terrain; only a glimmer of sunlight reflected from its crystalline shapes gives it away. The natural house, Frank Lloyd Wright wrote, should grow from its site, out of the ground and into the light "as dignified as a tree in the midst of nature." The Idaho home rises naturally from its mountain aerie, anchored by a base of native stone, its peaked glass rooflines, like flawlessly polished metamorphic rock, reaching toward the sky.

Entering the house is like entering a forest. Clusters of trees flank the steps to the formal entry. At the top of the steps is an airlock, or vestibule, lit by clerestory windows. The airlock insulates the house from blasts of cold alpine air. A second series of steps leads to sculpted glass doors. To either side are the living spaces. Straight ahead is the canopy of an indoor forest, a riverine ecosystem comprising a variety of terrestrial, wetland, and aquatic habitats. It is reached by a staircase that winds down to the forest floor, much like a canyon walk. The indoor ecosystem, complete with birds and fish, is the heart of the home and alludes to the plant communities that straddle the Big Wood River one hundred feet below.

Water, forest, earth, and sky become a part of the living environment. A snow-fed stream, one of several tributaries on the site, flows through the greenhouse. The water is pumped up and into the living room, where it cascades down rocks like a miniature waterfall, providing soothing sounds and humidifying the air with spray. Transparent bedroom roofs offer uninterrupted views of the nighttime sky; as they drift off to sleep the inhabitants can gaze into deep space, light years away. By day, window walls throughout the living spaces frame earthly mountain vistas.

The indoor wetlands and forest, however, are where the Hormel house truly breaks with convention and ushers in a new era of architecture. The riverine habitat is not just another living space boasting unique features but rather a living ecosystem. Human shelter merges with wild habitat; the apartheid is broken.

The indoor ecosystem promotes biodiversity and more. It adds

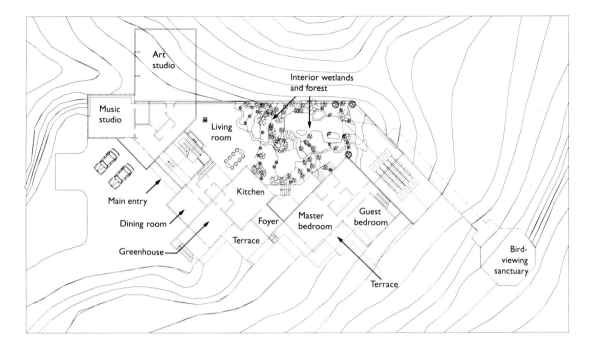

The plan labels: Art studio, Music studio, Living room, Interior wetlands and forest, Main entry, Dining room, Greenhouse, Kitchen, Foyer, Terrace, Master bedroom, Guest bedroom, Terrace, Bird-viewing sanctuary

▲
The house's indoor wetlands and forest are where it truly breaks with convention and ushers in a new era of architecture. Used water goes to an attached greenhouse where it is cleansed in a series of sun-bathed translucent cylinders, each an aquatic ecosystem with bacteria, snails, algae, and higher plants. The water then moves into man-made marshes where more pollutants are extracted. From there, it can flow into the indoor ecosystem, supporting fish, trees, and other plants.

◄
The interior landscape — complete with waterfalls, a river, a swimming pool, and trees and other lush tropical vegetation — is the heart of the 21st-century home. It adds beauty to the indoor environment, and repairs the ruptured link between nature and daily life.

beauty to the indoor environment, and repairs the ruptured link between nature and daily life — even in a high-elevation climate in which winter temperatures regularly plunge below zero. Just as important, it mitigates the adverse impacts of human activity on the fragile site. It cleanses the air of indoor pollutants while creating oxygen and adding humidity. It also purifies tainted water from the dishwasher and washing machine and all toilets, sinks, baths, and showers — constituting the first revolution in domestic plumbing since its arrival in the late nineteenth century.

In the typical house, tapwater comes from an underground well or nearby surface waters. Most of this drinking-quality water is misused. Fresh, clean water that comes out of the tap while we wait for hot water goes down the drain. Pure, potable water is routed to the toilet and becomes toilet water, which is referred to as blackwater. All of this wastewater, in addition to used water from the bathtub, sink, laundry, and shower (known as graywater), is then routed directly to the septic tank or municipal sewage treatment plant.

In the Ketchum house, drinking water comes from a well on site. Used water goes to an attached greenhouse. There, in a Living Technology system developed by biologist Dr. John Todd of Ocean Arks International in Falmouth, Massachusetts, it is cleansed in a series of sun-bathed translucent cylinders, each a functioning aquatic ecosystem with algae, bacteria, snails, and higher plants.

THE NEXT GENERATION OF SEWAGE TREATMENT

You've just flushed the toilet. Where does the sewage go? If you live in a rural area, the wastewater probably goes to a septic tank and underground leachfield. In most metropolitan areas, the water travels through municipal sewer pipes to a sewage treatment plant.

Conventional sewage treatment plants release into receiving waterways effluent (the fancy word for the liquid that leaves the facility) that's still full of nitrogen and phosphorus and toxic metals and chemicals. They also generate an estimated 8.5 million tons of sludge a year. Making further improvements in water quality with the mechanical and chemical technologies used in conventional advanced wastewater treatment would cost a fortune, create even more sludge, and in many cases require the use of substances that are far more dangerous then the pollutants they're supposed to remove.

Many home septic systems don't work well, either. One difficulty is "poor perc" — soils that can't percolate, or drain, fast enough to prevent wastewater from puddling up and creating a health hazard. In areas where the soil is extremely sandy, the problem is just the opposite: soils percolate all too well, and the wastewater reaches underground water supplies before the pollutants have been adequately broken down. What's more, sooner or later every septic tank needs to be pumped out, and the resulting septage must be treated and disposed of somewhere.

In a small but growing number of homes and communities, used water from the toilets, showers, sinks, laundry, and dish-

washer is being treated in bioengineered systems in which plants and animals break down the pollutants. These natural systems can *restore* water, not just treat it. They can do it relatively cheaply, and without generating massive quantities of sludge. One such system is Living Technology, developed by aquatic biologist Dr. John Todd. This system comprises a series of self-contained and self-regulating aquatic and wetland ecosystems (see page 71). Because warm-climate plants are used to purify the water, a Living Technology system is generally located in a greenhouse. One of the advantages of this technology is that it doesn't require much space; a household-size system fits easily in a small greenhouse. And because it is located in a greenhouse, it can be used even in the coldest climates.

Dr. Walter Adey's algal turf scrubber, described on pages 74–75, can treat well water and household graywater (used water from the shower, bathroom sink, laundry, and dishwasher) or squeeze out even more pollutants from water leaving a Living Technology or other bioengineered system. The water can then be used indoors or outdoors for bathing, washing clothes or dishes, to fill swimming pools or aquariums, or for irrigation.

At homes with more land, partially treated water flowing out of the septic tank is routed to man-made ponds and marshes, where wetland and aquatic flora and fauna purify the water.

The most affordable bioengineered treatment system is the microbial rock bed/plant filter developed by Dr. Bill Wolverton, formerly with NASA. These mini-marshes, which consist of cattails, rushes, and other wetland plants or even ornamental flowers, treat water from a septic tank and at the same time beautify the property. Experience with the home marshes in the southern United States indicates that the average two-bedroom house typically requires a 3 x 70 foot trench or a 20½ x 10¼ foot shallow sump. These are lined with an impermeable plastic membrane, which in turn is filled twelve inches deep with 1 to 1½-inch-round river rock or gravel and capped with 6 inches of pea gravel to cover the water, eliminate odors, and even allow for foot traffic. Plants are transplanted into the pea-gravel layer, with their roots reaching into the river rock zone. Pollutants in the waste-water are broken down by microorganisms and become food for the plants. Water leaving the system can be used to irrigate the garden. At $1,500 to $5,000, depending on local labor costs, microbial rock bed/plant filters usually cost less than the conventional septic tank and leaching field.

With the exception of John Todd's Living Technology, even these natural systems produce some sludge that must be removed every three years or so. Todd's system consumes most of the sludge.

TWENTY-FIRST-CENTURY BUILDING PRINCIPLES

- The home works with nature, not against it.
- The home is designed to minimize its contribution to global environmental problems such as ozone depletion, global warming, destruction of biodiversity, and air and water pollution.
- The home is transformed from an entity that *consumes* natural resources to one that *produces* or *restores* them. This is accomplished, for example, with photovoltaic cells that convert sunlight into electricity, solar hot-water heating, ground-source heat pumps for heating and cooling, water conservation, reclamation and recycling, and year-round food production in outdoor organic gardens and indoor hydroponic gardens and aquaculture systems.
- The home is designed to conserve resources. It uses 75 percent less energy than today's conventional dwellings and restores and reuses all water. Fifty percent of the material used for construction comes from recycled or salvaged sources and 90 percent is easily recyclable. Construction wastes are salvaged and reused.
- To support sustainable forestry practices in both the United States and the tropics, the home incorporates at least one domestic or rainforest wood that is certified by Scientific Certification Systems, the Rainforest Alliance, or other reputable organization.
- The home is constructed with at least one salvaged wood to support the efforts of small-scale lumber companies in the United States that recycle forest products that would otherwise be disposed of in landfills or burned in incinerators.
- All home appliances are free of ozone-depleting chemicals.

- Toxic and polluting petrochemical-based building materials, furnishings, pest controls, and cleaning products are avoided.
- The home promotes human health. Exposure to harmful chemical products, noise, electromagnetic fields, radon, and air and water pollution is minimized. Organic food production and physical exercise are encouraged.
- Every effort is made to preserve and promote biodiversity at the building site. Great care is taken to disturb as little as possible of the native vegetation during construction. Where the natural vegetation is long gone, native plant communities are restored. The landscape is designed to promote natural heating and cooling and enhance wildlife habitat. Wild ecosystems treat sewage and perform human life-support functions inside the home as well.
- The life-cycle cost (purchase price plus cost to operate and maintain) of each building system is paramount. Building components are not viewed as short-term investments.
- To reduce heating and cooling loads, the home design makes the most of solar orientation, thermal mass, natural ventilation, and shading.
- The home has a comprehensive recycling system to reduce by 75 percent the generation of household discards, including metals, glass, plastics, paper, and food and garden trimmings.
- The home is designed for several generations — eighty to one hundred years at a minimum.
- The home is beautiful. It complements the natural surroundings and is an expression of human imagination and craftsmanship.

The various organisms in these self-sustaining ecosystems feed on the waste and are eaten in turn by the larger creatures. From here, the water moves into man-made marshes composed of coarse sand and gravel and planted primarily with bulrush, where more pollutants are extracted.

From the Living Technology system, the water can be diverted to numerous other uses or further treatments. It can be piped to holding tanks for outdoor irrigation and backup fire suppression. It can flow into the interior riverine ecosystem, supporting fish, trees, and other plants. It can be piped to tanks in which fish are farmed and/or plants are grown hydroponically. It can be returned to the toilets for waste removal. It can also be directed to another living water-treatment system, an algal turf scrubber, and completely restored to its original pristine state. The restored water can once again be used for bathing, swimming, dishwashing, and clothes washing, and if it is subjected to UV rays to destroy potential viruses, even for drinking. In the algal turf scrubber, developed by the Smithsonian's Dr. Walter Adey, a counterbalanced, table-

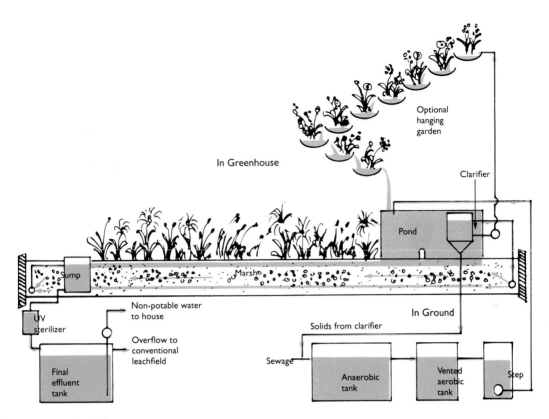

In Greenhouse

Optional
hanging
garden

Clarifier

Pond

Marsh

In Ground

Sump

Non-potable water
to house

UV
sterilizer

Solids from clarifier

Overflow to
conventional
leachfield

Sewage

Final
effluent
tank

Anaerobic
tank

Vented
aerobic
tank

Step

In this home-size Living Technology treatment system, sewage goes to underground tanks where bacteria begin breaking down pollutants. The wastewater then flows into a greenhouse, where fauna and flora in an indoor pond continue the cleansing process. From there, the partially treated water can go to an optional hanging garden, or directly to a man-made marsh, where most of the rest of the pollutants are extracted. After passing through a sterilizer where it is subjected to UV rays to kill any remaining viruses, the treated water can go back into the house for recycling.

sized trough fills with the water and then tips onto a mesh on which algae grows. The algae consumes the remaining pollutants. At River's Edge no contaminated water leaves the site, and all water is recycled, used two or three times and sometimes more.

The Idaho house is self-sufficient, or close to it, in other ways as well. Like the other houses featured in this section, it conserves precious fossil fuels. In fact, homes of the twenty-first century must be net energy producers, not consumers of energy. The Idaho home is designed to achieve this goal with a variety of energy sources: for example, the house captures solar heat passively by its south-facing windows. Its mass walls store the heat and radiate it to the living spaces later in the day, when temperatures plummet. Its photovoltaic system converts sunlight into electricity. Wind energy is used to pump water and generate additional electricity. Primary heating and cooling are provided by a ground-source heat pump that extracts heat from a well in winter to warm the rooms and cools them in summer by transferring heat from the home to the underground water. Daylight illuminates the living spaces, reducing the amount of electricity needed for artificial lighting.

The house produces other resources as well. Herbs, salad greens, and selected vegetables are produced in the kitchen greenhouse year-round. Additional fruits, nuts, and vegetables are grown in the interior forest and outdoors in a high-desert garden.

The house is constructed with a minimum of material. Each material serves a function, in most instances more than one function, and there are few extraneous details. Glass is employed dramatically throughout the home. Walls, floors, staircases, and ceilings are made of lightweight concrete. Both aerated concrete and glass are made from natural minerals in plentiful supply; they are also chemically inert and therefore do not pollute the indoor air. Woods from environmentally responsible sources are used in a few locations to full effect — in flooring, cabinetry, and furniture. For example, parquet flooring is made from remilled shipping pallets. However, most floors are stone, with plant-fiber carpets and cotton and wool rugs used for warmth and color. But by far the most striking materials are the terrestrial, wetland, and aquatic habitats that merge with the man-made technologies, forming the house's core.

At the turn of the twentieth century, Frank Lloyd Wright declared that the natural house would help liberate Americans from the decadent traditions of Europe by awakening in us "the desire for such far-reaching simplicities of life as we may see in the clear countenance of nature." So that we could more clearly see nature, he merged the inside and the outside through expanses of glass, and further fused building and landscape by using stones and woods indigenous to the area of construction. But the union of nature and architecture envisioned by Wright was ultimately aesthetic and theoretical; the natural house was designed so that people could view nature passively through transparent walls. The inhabitants were outsiders, mere observers of nature, not "citizens of the land community" to use the words of conservationist and author Aldo Leopold. By contrast, at River's Edge, nature permeates the structure. The living systems recreated in the home support human life. In the twenty-first-century home, we nurture nature, as nature nurtures us. The delicate and urgent task of balancing human and ecological needs becomes a part of daily life.

PART THREE

COMPENDIUM OF

ENVIRONMENTAL HOME

BUILDING, DECORATING, AND

MAINTENANCE PRODUCTS

WHAT MAKES A MATERIAL ENVIRONMENTAL?

When architects and interior designers choose materials for a home, they consider a number of factors, including how the materials will look, how much they will cost, how easy they are to find and install, and whether they're durable and easy to maintain. Until recently, however, no one considered the environmental and health impacts of building and decorating materials.

One reason for this omission is that for decades the building industry, as well as homeowners, had access to a seemingly unlimited supply of high-quality woods and other materials and inexpensive supplies of energy. Another reason is that so-called life-cycle studies of the environmental "costs" of producing, shipping, installing, and maintaining these materials have been under way only for the past several years. However, more data is available every day, making the task of comparing materials ever easier.

The following checklist will give you some idea of the range of environmental impacts associated with every material used in the home. These are the kinds of questions you should ask whenever you shop for a home building, decorating, or maintenance product.

Raw Materials Acquisition (Mining, Harvesting, Drilling)

- Is the raw material in limited supply, or can it be replaced by nature by the time its useful life as a building material is up? As an example, compare aluminum- and wooden-frame windows. Bauxite, from which aluminum is made, comes from dwindling deposits in the tropics. By contrast, a wooden window made with a tree species in plentiful supply, if well constructed, will last the fifty or so years it takes for another tree to grow. (It will also need to be maintained regularly, preferably with low-pollutant paints or finishes.)

- How much energy is needed to mine, harvest, or drill the material? Is the energy provided by nonrenewable fossil fuels?

- How is the surrounding environment affected? For example, are forests or disappearing ecosystems or wildlife habitats destroyed?

Raw Materials Processing and Manufacturing

- How much energy is consumed during manufacturing? What kind? For example, the aluminum in the example above is extremely energy intensive to produce, unlike the wood, which requires little fuel to mill.

- How much water is consumed during manufacturing?
- How much air and water pollution is created?

Product Packaging

- Is the packaging made of recycled materials? Can it be recycled?
- Is the packaging made with materials that are harmful to the environment, such as polystyrene foam made with ozone-depleting chemicals?

Product Distribution

- Must the material be shipped long distances or is it produced locally, saving energy that would otherwise be used to transport it?

Product Installation, Use, and Maintenance

- Does installation, use, or maintenance result in pollution that threatens your health or that of workers you may hire?
- Does the product contribute to your home's energy efficiency? Water efficiency?
- How durable is the product? Will it last a long time, or will it need to be replaced relatively quickly?

Disposal, Reuse, or Recycling

- Is the product made with recycled material?
- Can it be reused?
- Can it be recycled?
- Must the product be disposed of in a landfill specially designed to contain hazardous waste?

From the environmental point of view, there is no "perfect" building material. The perfect product would be made of 100 percent renewable materials. Fossil fuels would not be needed to harvest, drill, or extract it, or to manufacture, transport, or operate it. The product would not pollute the air inside your home, and it would present no health threat to manufacturers and installers. What's more, it would be possible to reclaim and recycle the product indefinitely.

While there are no perfect products, there certainly are environmentally preferable ones. The products recommended in this book have been chosen because they meet five major criteria:

1. They do not pollute the air inside your house, or at least produce less indoor pollution than conventional products.
2. They conserve natural resources. Most importantly, buying them does not add to the pressure on threatened species or ecosystems or deplete protective ozone in the upper atmosphere.
3. They are water conserving and energy efficient.
4. Mining and manufacturing them does not result in excessive air and water pollution.
5. They are recyclable or biodegradable.

A house designed and constructed with these materials will be beautiful. It certainly will be healthier to live in than the typical home. It will also be more comfortable. And because it is suited to the local climate and designed to last, it will be a lot less expensive to maintain and to run.

HOW TO USE
THE COMPENDIUM

The compendium is divided into seven sections, according to where and how the various materials are used in the home:

> The Structure
> Life-Support Systems
> Interior Surfaces
> Furnishings
> Appliances and Fixtures
> Children's Products
> Home Maintenance Products

The discussion of each building and decorating material is divided into several sections, beginning with a description of conventional products and the major health and environmental concerns associated with them. Next, the environmental alternatives are explored. Last is a suppliers section, in which specific products are recommended. The names of the manufacturers and their addresses and phone numbers are included so that you can write or call for additional information, to place an order, or to obtain the names and locations of the nearest distributors. For additional information on the health effects of the major indoor pollutants, see Appendix A, "A Guide to Home Pollutants." For a look at how environmental products and technologies work together in the home, in both renovations and new construction, see "Building, Remodeling, and Decorating with Environmental Materials," beginning on page 31.

THE STRUCTURE

For years when environmental builders used the word "radiation," they were referring to *solar* radiation and how to capture it to heat a home. That changed in 1984, when Stan Watras, an engineer working on the Limerick nuclear power plant in Pottstown, Pennsylvania, began setting off radiation detectors at the generating station. Subsequent tests revealed that radioactive radon gas in his home, not radiation at the power plant, was the source of contamination. In fact, radon levels in his house were so high that he and his family were forced to relocate. Since 1984, scientists and builders have learned a lot about radon contamination of homes, and what to do about it.

Radon, a naturally occurring gas emitted by the earth, has always been present in human shelters. However, it is emitted in far greater quantities from certain types of bedrocks and soils (see the map of estimated radon levels in the United States on page 82). The tightly constructed buildings resulting from concern for energy efficiency in recent years trap radon that seeps into homes, exacerbating the problem.

Radon is a colorless, odorless gas produced when uranium in the earth's crust decays. Radon itself decays further into polonium and eventually into a stable isotope of lead. As it decays, it emits alpha particles and other forms of radiation, which can cause chemical changes as they pass through the human body.

Because radon is a gas, it can seep into your house through cracks in the foundation, sump holes, gaps between floors and walls, and openings for plumbing in the basement walls and floor. Two short-lived decay products of radon, polonium-218 and polonium-214, both solids, tend to attach themselves to dust and other particles in the air that are easily inhaled. The contaminated dust can lodge in your lungs, where alpha particles emitted by the polonium can cause mutations, and sometimes cancer.

Radon can enter your home through water as well as through the soil. If your water comes from a well, it may have absorbed some radon from rocks underground. Most public water supplies, which generally come from rivers, lakes, and other surface waters, have very low concentrations of radon because much of the gas escapes when it is exposed to air. The principal known health effect from radon in domestic water is lung cancer, which may result from breathing radon released from the water into indoor air, especially in the shower. There is a lesser risk of stomach cancer from ingesting it.

The U.S. Environmental Protection Agency

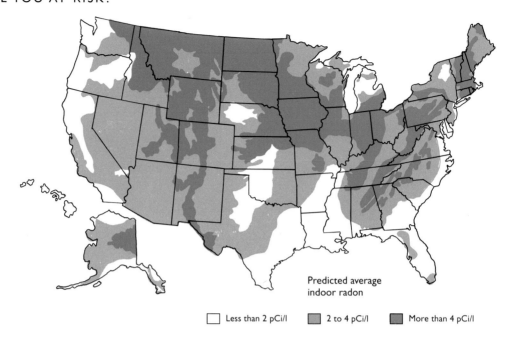

Predicted average
indoor radon

☐ Less than 2 pCi/l ■ 2 to 4 pCi/l ■ More than 4 pCi/l

This map, produced by federal and state agencies, shows the potential for indoor radon problems in each state. The Environmental Protection Agency recommends taking action to lower levels that are 4 picocuries per liter (pCi/l) or higher.

currently recommends that you take action to reduce the amount of radon in your home if levels are above 4 picocuries per liter (pCi/l) in the air. (Picocuries per liter is the standard unit of measurement for radioactive radon. Some Scandinavian countries have set a much more stringent limit of 2pCi/l in indoor air.) The EPA estimates that about 6 percent of American homes have radon levels of 4pCi/l or more.

The EPA has proposed a standard of 300 pCi/l of radon in public drinking water. Radon levels in private wells typically exceed this level. The highest levels have been found in the Northeast.

Health Hazards and Environmental Concerns

Medical researchers are still arguing about the magnitude of the health risk posed by radon. Based on studies of miners who were exposed to large amounts of radon on the job, national and international committees of experts calculate that 1 to 3 percent of people exposed to 4 pCi/l for their entire lives will die of lung cancer caused by radon. The Environmental Protection Agency estimates that radon is the primary cause of 10 percent of lung cancer cases in this country. This is an average rate that includes both smokers and nonsmokers. If you smoke, your risk is higher; if you don't, it is lower. Risk assessment is not a precise science. However, according to the best information available, the EPA and the Surgeon General currently believe that exposure to radon in buildings causes between five thousand and twenty thousand cases of lung cancer every year.

HOW RADON GETS INTO YOUR HOUSE

Because radon is a gas, it can seep into your house through cracks in the foundation, sump holes, gaps between floors and walls, and openings for plumbing in the basement walls and floor. It can also come from certain stones used to build and decorate your home. Radon can contaminate your indoor air through tapwater as well as through soil, particularly if your water comes from a well.

The good news is that the data from research on underground miners indicates that after they left their high-radon work environment their high lung cancer risk declined. Their bodies seemed to recover from the radiation. This suggests that if you test for radon in your home and take steps to lower levels if necessary, you can not only avoid future health damage but also begin to undo damage already done.

Environmental Products

Although the lung cancer risk is alarming, much research has been done during the past decade that has made it possible to detect radon quickly and remedy the problem relatively inexpensively or, better yet, design new construction to prevent radon from ever becoming a health problem in your home. From radon detectors to radon-blocking sealants, an array of new building products are now available.

What to Do About Radon

- Contact your regional Environmental Protection Agency office and your state radon hotline for more information about radon risks in your area. Phone numbers and addresses are listed on pages 85–86.

- If you're building a new house in a risk area for radon, build in radon protection. Because effects of low-level radiation such as radon in indoor air are poorly understood, and indeed some countries set maximum allowable levels lower than Environmental Protection Agency recommendations, consider building in even more protection than is currently recommended by the agency (see "A Radon Prevention System for New Homes," page 89).

- If your home is in an area at risk for radon, test the air for potential problems. Because you can't see or smell radon, special equipment is needed to detect it. The two most popular radon detectors are the charcoal canister and the alpha track detector. The former is used for short-term, or screening, tests that can indicate whether you have a potential radon problem. However, recommended maximum radon levels are based on annual averages, not one or two short-term tests, and so the EPA recommends that you use a short-term test only to screen for potentially high radon levels. Conduct four short-term tests for best results. If there is any indication that levels in your home are high, you should take a long-term test.

  The alpha track detector contains a piece of film called an alpha track, which is placed in your home for a period of three months to a year. Alpha particles, emitted when radon decays, leave marks in the film. When the test is over, technicians count the tracks and estimate the amount of radon in your home. Because this test is conducted over a longer period, it represents your average radon exposure. For the best assessment of radon levels in your home, conduct two long-term tests.

  Be sure to test those rooms that you and your family spend the most time in — bedrooms, kitchen, and family room, for example. And don't forget the nursery; although there are no studies of children exposed to radon to determine whether they're more sensitive than adults, some research on other types of radiation exposure indicate that children may be more vulnerable and therefore more at risk. For more information on how to test, call your regional Environmental Protection Agency office for a copy of the EPA booklet *A Citizen's Guide to Radon: What It Is and What to Do About It.*

- If you're renting a house or an apartment on the ground or first floor, you should also test for radon. The number of times you should test depends on how long you'll be there: short-term charcoal canister tests are probably sufficient for short-term rentals; long-term alpha track tests are recommended for long-term rentals of two years or more.

- Take corrective actions if necessary. If your test results are 200 picocuries per liter or higher, the EPA recommends that you take immediate steps to reduce levels as much as possible or relocate until radon levels can be reduced. Exposures in the range of 20 to 200pCi/l are far above average and the agency recommends that you take action within several months. If your results are about 4pCi/l to 20pCi/l, it advises that you take action within a few years at most.

FOR MORE INFORMATION

STATE RADON CONTACTS

Alabama	(800) 582-1866	Montana	(406) 444-3671
Alaska	(800) 478-4845	Nebraska	(800) 334-9491
Arizona	(602) 255-4845	Nevada	(702) 687-5394
Arkansas	(501) 661-2301	New Hampshire	(800) 852-3345 x4674
California	(800) 745-7236	New Jersey	(800) 648-0394
Colorado	(800) 846-3986	New Mexico	(505) 827-4300
Connecticut	(203) 566-3122	New York	(800) 458-1158
Delaware	(800) 554-4636	North Carolina	(919)571-4141
District of Columbia	(202) 727-5728	North Dakota	(701) 221-5188
Florida	(800) 543-8279	Ohio	(800) 523-4439
Georgia	(800) 745-0037	Oklahoma	(405) 271-5221
Hawaii	(808) 586-4700	Oregon	(503) 731-4014
Idaho	(800) 445-8647	Pennsylvania	(800) 237-2366
Illinois	(800) 325-1245	Puerto Rico	(809) 767-3563
Indiana	(800) 272-9723	Rhode Island	(401) 277-2438
Iowa	(800) 383-5992	South Carolina	(800) 768-0362
Kansas	(913) 296-1560	South Dakota	(605) 773-3351
Kentucky	(502) 564-3700	Tennessee	(800) 232-1139
Louisiana	(800) 256-2494	Texas	(512) 834-6688
Maine	(800) 232-0842	Utah	(801) 538-6734
Maryland	(800) 872-3666	Vermont	(800) 640-0601
Massachusetts	(413) 586-7525	Virginia	(800) 468-0138
Michigan	(517) 335-8190	Washington	(800) 323-9727
Minnesota	(800) 798-9050	West Virginia	(800) 922-1255
Mississippi	(800) 626-7739	Wisconsin	(608) 267-4795
Missouri	(800) 669-7236	Wyoming	(800) 458-5847

FEDERAL RADON CONTACTS

■ Regional Environmental Protection Agency Offices

State — EPA Region

Alabama — 4	Connecticut — 1	Hawaii — 9	Kentucky — 4
Alaska — 10	Delaware — 3	Idaho — 10	Louisiana — 6
Arizona — 9	District of	Illinois — 5	Maine — 1
Arkansas — 6	Columbia — 3	Indiana — 5	Maryland — 3
California — 9	Florida — 4	Iowa — 7	Massachusetts — 1
Colorado — 8	Georgia — 4	Kansas — 7	Michigan — 5

Minnesota — 5	New Jersey — 2	Oregon — 10	Utah — 8
Mississippi — 4	New Mexico — 6	Pennsylvania — 3	Vermont — 1
Missouri — 7	New York — 2	Rhode Island — 1	Virginia — 3
Montana — 8	North Carolina — 4	South Carolina — 4	Washington — 10
Nebraska — 7	North Dakota — 8	South Dakota — 8	West Virginia — 3
Nevada — 9	Ohio — 5	Tennessee — 4	Wisconsin — 5
New Hampshire — 1	Oklahoma — 6	Texas — 6	Wyoming — 8

EPA Regional Offices

EPA Region 1
Room 2203
JFK Federal Building
Boston, MA 02203
(617) 223-4845

EPA Region 2
26 Federal Plaza
New York, NY 10278
(212) 264-2515

EPA Region 3
841 Chestnut Street
Philadelphia, PA 19107
(215) 597-8320

EPA Region 4
345 Courtland Street, NE
Atlanta, GA 30365
(404) 881-3776

EPA Region 5
230 South Dearborn Street
Chicago, IL 60604
(312) 353-2205

EPA Region 6
1201 Elm Street
Dallas, TX 75270
(214) 767-2630

EPA Region 7
726 Minnesota Avenue
Kansas City, KS 66101
(913) 236-2803

EPA Region 8
Suite 1300
One Denver Place
999 18th Street
Denver, CO 80202
(303) 283-1710

EPA Region 9
215 Fremont Street
San Francisco, CA 94105
(415) 974-8076

EPA Region 10
1200 Sixth Avenue
Seattle, WA 98101
(206) 442-7660

Okay, your tests have indicated that radon levels should be reduced. Where do you go from here?

The first thing to do is to consult with regional and state radon authorities and local radon-mitigation contractors. Identify the most likely sources of radon in your home. Determine what kinds of remediation are most appropriate for your particular situation. Then figure out which option or combination of options will reduce levels and health risks most economically. Radon mitigation runs the gamut from simply opening windows to installing mechanical ventilation systems such as fans or heat-recovery ventilators. Other courses of action include sealing the foundation, installing a sump pump liner and/or lid, sealing or removing building materials that contain radon, and using a high-efficiency particulate (HEPA) filter, which to some extent can trap radon-affixed particles present in your indoor air.

- If your water comes from a well, test it for radon levels. Of primary concern are levels high enough to release into your air, from everyday water use, 4 pCi/l or more. Every 10,000 pCi/l of radon in water contributes about 1pCi/l of radon to household air.

- Should radon levels in your water exceed 40,000 pCi/l, you should definitely take action, because this will cause levels in your household air to exceed the 4pCi/l guideline. If your water radon level is between 5,000 and 40,000 pCi/l, there is still a significant risk of lung cancer from breathing the radon released from the water. Levels below 5,000 pCi/l, particularly below 1,000 pCi/l, pose less of a lung-cancer risk, and the cost-effectiveness of treating the water should be evaluated on a case-by-case basis. Two types of water filtration systems can help bring radon levels down: activated carbon and spray or bubble aeration.

Suppliers

RADON DETECTORS
Professional and licensed testing consultants can be retained to conduct tests and make recommendations. Fees will range from $200 to $500. However, you can take samples of air and water yourself.

AIR SAMPLING
Activated charcoal, the most commonly used type of testing device, is suitable for short-term tests only, from three to seven days. The cost is $10 to $25 per test. Alpha track radon detectors range in cost from about $25 to $50.

Air-Check Sampler, Air Chek, Inc.
P.O. Box 2000, 180 Glenbridge Road, Arden, NC 28704; (800) 247-2435 or (704) 684-0893

Canister Pak, Teledyne Isotopes
50 Van Buren Avenue, Westwood, NJ 06775; (800) 666-0222

DMA Radtech, Inc.
1 Research Circle, Waverly, NY 14892-1532; (607) 565-3500

E-Perm Detector, REEP, Inc.
300 Corporate Court, South Plainfield, NJ 07080; (800) REEP-INC

Quick Screen and RadTrak, Tech/Ops Landauer, Inc.
2 Science Road, Glenwood, IL 60425; (800) 528-8327 or (312) 755-7000

Radex Kit and Monitor, Ecodex
1 Aaron Road, Lexington, MA 02173; (800) 4-ECODEX, or (617) 862-4300

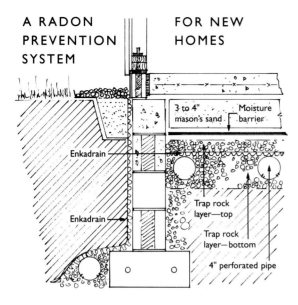

A RADON PREVENTION SYSTEM

FOR NEW HOMES

3 to 4" mason's sand

Moisture barrier

Enkadrain

Enkadrain

Trap rock layer—top

Trap rock layer—bottom

4" perforated pipe

Radon Testing Corporation of America
Trent Building, P.O. Box 258, Irvington, NY 10533; (800) 457-2366

WATER SAMPLING
Radlab
4915 Prospectus Drive, Suite C1, Durham, NC 27713; (919) 544-9080

CONTINUOUS RADON MONITORS
Continuous home radon monitors are another option. They are available in both desktop and permanently installed models. Readings are constantly displayed, and indicator lights signal low, medium, and high readings. Such systems will also activate or deactivate dampers, fans, and blowers as necessary. They cost from about $250 to $300.

Home Radon Monitor, Honeywell, Inc.
Honeywell Plaza, Minneapolis, MN 55408; (800) 345-6770

FOUNDATION SEALANTS
AFM Enterprises, Inc.
1140 Stacy Court, Riverside, CA 92507; (714) 781-6860
Three products are available: Dyno Flex is a paste for repairing cracks or gaps. Dyno Seal is a liquid sealant that will adhere to concrete or wood and is excellent for basements. Penetrating Water Seal is designed for concrete, porous bricks, tile, and stone.

Safe-Aire
162 East Chestnut Street, Canton, IL 61520; (800) 331-2943
An entire line of radon mitigation products, including liquid radon sealant.

Skanvahr Coatings
18646 142nd Avenue N.E., Woodinville, WA 98072; (206) 487-1500
Skanvahr EF, suitable for sealing wood, slate and stone, cement, and brick.

FANS
Fantech
1712 Northgate Boulevard, Sarasota, FL 34234; (813) 351-2947

Infiltec
P.O. Box 8007, Falls Church, VA 22041; (703) 820-7696

Safe-Aire
162 East Chestnut Street, Canton, IL 61520; (800) 331-2943

See also Heat-Recovery Ventilators in "Mechanical Ventilation," pages 154–157, and HEPA Air Filtration Units in "Air Filtration and Purification," pages 154–159.

A RADON PREVENTION SYSTEM FOR NEW HOMES

Because the effects of low-level radiation are not fully understood, and because some countries recommend radon levels well below the 4pCi/l American standard, you may want to consider building in more protection than that afforded by the variety of sub-slab radon venting techniques currently recommended by the U.S. Environmental Protection Agency. The following system is used routinely by the Masters Corporation, Paul Bierman-Lytle's design and building firm. It consists of a gravel bed over compacted or undisturbed soil with a network of perforated pipes covered by Enkavent, a moisture barrier, sand, and the finished slab (see the illustration on page 88).

Perforated pipes four inches in diameter (usually PVC pipe, available at most lumberyards) are installed around the entire inner perimeter of the slab below the top of the footings, with the holes in the pipe facing downward. These pipes are pitched on a slope and either pass through or under the footings to a "radon house" (a tiny outbuilding erected at least fifty feet away from the house) or drywell from which radon can be vented to the outside air. Additional pipes are installed under the slab, usually five to eight feet on center, connecting with the perimeter pipes. At various locations, "tee" connectors are installed with caps above the final slab height to make it easy to do radon testing and flush out the system with water periodically. As an alternative to a radon house, vertical ducts attached to the tee connectors can vent the radon out of the roof.

The sub-slab pipes are covered with ¾-inch-diameter stone or trap rock either twelve inches deep or to the height of the footing. The stone is covered in turn with Enkavent or other moisture barrier to keep the concrete slab from clogging up the radon air bed. Sand is laid above the groundsheet to protect this crucial barrier from rips or punctures during the installation of the concrete slab.

Because the pipes are pitched ¼ inch to the foot in order to move any possible groundwater away from the slab, the system also facilitates drainage and keeps the slab moisture-free.

Once construction is complete, radon testing begins. If levels are high, mechanical fans can be installed at the radon house, the drywell, or the metal ducts inside the structure. In high-density areas, entire blocks of houses or neighborhoods can be hooked to a central radon house.

Cost of a single-house system: $1,000 to $1,500.

IMPERMEABLE BARRIERS

These products create a barrier to prevent seepage of radon from the soil into your home, and create a uniform blanket of airspace from which to intercept radon gas under your basement or slab so that it can be safely vented. They usually are installed in new homes before the slabs are poured.

Enkadrain and Enkavent, American Enka Company

Enka, NC 28728; (704) 665-5050

Enkadrain is a foundation waterproofing membrane. Enkavent is used under the slab to create an air blanket from which radon can be vented. Enka-B, a similar product, can be used along the basement walls for additional protection.

New England Liner

35 Wooster Court, Bristol, CT 06801; (203) 583-5463

A non-tearing vapor barrier for use under slabs.

SUB-SLAB RADON COLLECTORS

These are prefabricated boxes or plenums used to transport radon away from the foundation.

Safe-Aire

162 East Chestnut Street, Canton, IL 61520; (800) 331-2943

SUMP PUMP LINERS AND LIDS

Lining your sump pump chamber and re-covering it with a tightly fitting lid can help reduce your exposure to radon.

Safe-Aire

162 East Chestnut Street, Canton, IL 61520; (800) 331-2943

RADON REMOVAL SYSTEMS FOR WELL WATER

The Stripper, Lowry Aeration Systems

4915 Prospectus Drive, Suite C1, Durham, NC 27713; (919) 544-9080

A bubble aeration system in which radon is released from the water to the air bubbles and is then vented outside.

CONCRETE

The use of concrete in residential construction dates back at least two thousand years to ancient Rome. The ruins of Emperor Hadrian's villa near Rome reveal that some walls and domes were constructed with a mixture of volcanic tuff and sand very much like modern concrete.

The development of portland cement in 1824 by an English inventor led to the first major improvement in concrete mortar. This binding agent is the major component of modern structural concrete. Reinforced concrete (concrete in which metal bars have been placed to increase strength) was developed later in the nineteenth century, and pre-stressed concrete (pre-cast concrete reinforced by steel cables placed under great tension) followed early in this century. What began as a mortar was transformed into complete structural components, from the concrete blocks used in home foundations to the huge columns and piers that support skyscrapers and bridges.

It is concrete's remarkable strength and its ability to be formed into a multitude of shapes that make it one of the most widely used construction materials today. It can be pre-cast under ideal factory conditions and shipped to the building site as slabs, beams, columns, sculpture, furniture, pipes, and countless other components. It can also be cast in place at the construction site. It can be poured between forms and allowed to set, or sprayed on as a finish. Concrete has become so mainstream, in fact, that it's hard to envision a home that does not incorporate it in myriad forms, from footings to sidewalks to septic tank to gutter splash basins. In modern solar houses it is used as a thermal mass in walls and floors to retain the sun's heat. Concrete is completely resistant to fire and rot and does not require protective coatings. It lasts so long that it is essentially permanent, as ancient Roman ruins attest — in some cases, the concrete mortar between bricks and stones survives, while the bricks and stones themselves are long gone.

Conventional Products

Concrete is comprised of three major components: aggregates, usually a mixture of coarse and fine materials such as gravel, crushed stone, and sand; portland cement; and water. Portland cement is made from lime, silica, iron, and alumina. The water reacts chemically with the portland cement during a curing process to bind the materials together. A variety of additives are mixed in to improve curing, make the concrete more resistant to damage from freezing, improve pouring and mixing, and even add color.

Health and Environmental Concerns

Mining

Most of the raw ingredients of concrete must be mined, including the limestone, silica, iron, and alumina that make up portland cement, as well as the crushed rock or gravel used as aggregates. Some of these materials are taken from surface mines, which have a number of deleterious environmental effects: topsoil is disturbed, erosion occurs, and the quality of surrounding ground and surface waters can suffer. Coal and iron mining can be particularly harmful, as acid runoff, or leachate, sulfates, and toxic metals resulting from these operations can pollute surrounding waters. When materials are mined from particularly fragile environments — alumina from tropical forests, for example — damage can be even worse.

Manufacturing Wastes

Portland cement, the most important component of concrete, not only requires substantial energy to manufacture but also results in considerable air pollution. After they are mixed together, the lime, silica, iron, and alumina in portland cement are heated to 2,900 degrees F in a rotary kiln, usually fired by gas, coal, or oil, to make a pellet-like material called clinker. After cooling, the clinker is ground into a powder and mixed with gypsum. State-of-the-art technologies required by federal and state environmental officials effectively control many emissions from the kilns. However, the carbon dioxide produced by the combustion of the fuels, as well as oxidation of the lime, represents a significant addition of CO_2 to the atmosphere, according to the World Resources Institute, an environmental research organization based in Washington, D.C. In 1987 in the United States, for example, 9.8 million metric tons of CO_2 were emitted as a result of the manufacture of some 76 million metric tons of concrete. Many scientists have linked increases of carbon dioxide in the atmosphere, along with other so-called greenhouse gases, to potential global warming.

Solid Waste

It is estimated that concrete accounts for 67 percent of the weight and 53 percent of the volume of construction demolition wastes in North America. In the United States, this amounts to a significant part of the larger waste stream. Yet very little concrete — probably less than 5 percent — is recycled.

Environmental Products

Concrete, particularly the new lightweight versions, is destined to be one of the most important components in the twenty-first-century house. As wood and other building materials become scarcer and more expensive, the new forms of concrete grow ever more attractive. The home of the future will no doubt be built of fewer of the construction assemblies that are so common today. Consider walls, for example. From outside to inside they consist of: two or three layers of paint or other coating over siding, an air infiltration barrier fixed to structural members like wood or steel, which hold the electrical and plumbing systems, as well as insulation, a vapor barrier, gypsum wallboard, and finally two or three coats of paint or wallpaper and paste.

Some of the new variations on concrete offer an all-in-one alternative to these complex assembly systems, much like traditional adobe brick. They have good insulating properties and structural strength and don't burn or rot easily. Even without sealers or other finishes, they're durable and long-lasting. They're lightweight and easily worked with tools. What's more, they can be recycled or made from recycled materials. And they're good-looking.

A number of new products are already available that are excellent alternatives to conventional concrete, including products for use in foundations and full basements, major users of concrete in the contemporary American home. Some are made with recycled by-products of other industrial processes, reducing the nation's solid waste burden.

In addition to these new products, modern adaptations of indigenous building systems such as adobe offer an environmentally sound alternative to concrete in some parts of the country. They require much less energy to manufacture and are simple to construct. And these resource-conserving technologies use a minimum of wood, an increasingly scarce commodity.

Suppliers

AUTOCLAVED AERATED CONCRETE

Autoclaved aerated concrete (AAC), developed by a Swedish architect in 1928 and utilized in other industrialized countries for years, is now becoming available in the United States and Canada. Also known as autoclaved cellular concrete, it is produced by adding small amounts of aluminum powder to a standard cement mix. The aluminum reacts chemically to create millions of tiny hydrogen bubbles within the concrete, expanding the material to about twice its original volume. As the concrete cures, the hydrogen dissipates and the pockets fill with air. Once the expansion is complete, the material is cut into blocks, slabs, or other shapes and moved into an autoclave, an airtight chamber that is filled with pressurized steam. The ten- to twelve-hour autoclaving process causes a second chemical reaction that gives the highly porous concrete its strength and durability.

With its high air content, AAC can weigh two thirds less than conventional concrete and has insulation values of up to R-1.6 per inch — more than twice that of old-fashioned concrete. The pre-cast material is comparable in strength to its conventional counterparts, yet its lower weight also makes it easier to transport, lift, and assemble.

Because it is lightweight and easily worked with carpenters' tools, autoclaved aerated concrete is sometimes compared to wood. Unlike wood, however, it resists rot and does not burn. It can be painted, plastered, or covered with wood paneling or tiles, or it can be left unsealed and unpainted. It currently costs 10 to 15 percent more than ordinary concrete. Yet when the reduction in costs associated with insulation, transportation, handling, assembly, and coatings or other finishes are factored in, it is quite competitive with conventional building systems.

The aggregate used to make AAC can be either sand or waste fly ash from coal-burning power plants. In 1992 approximately 45.5 million tons of fly ash were produced in this country. Only about a quarter of the fly ash produced annually is reused. Fly ash can constitute as much as 70 percent of AAC, so the potential for recycling is huge. The downside to this material is that there is little data on the potential environmental hazards of fly ash, although it is commonly used in AAC plants abroad.

Hebel U.S.A.
2305 Six Branches Drive, Roswell, GA 30076; (404) 552-8665

North American Cellular Concrete
3 Regency Plaza, Suite 6, Providence, RI 02903; (401) 621-8108

FOAM CRETE

Foam Crete, or more technically polymer cellular concrete, is a lightweight aerated concrete in which plastic foam is used in place of aggregates. It was developed by polymer chemist Gary Zeller.

Unlike AAC, Foam Crete can be prepared on site. It weighs one third less than conventional concrete yet has good compressive strength and can be used for structural as well as nonstructural support. Because it is prepared at the building site, transportation costs are minimal. The manufacturer claims that Foam Crete contains no toxic ingredients and releases no hazardous vapors. It can be recycled, although it contains no recycled material.

Like AAC, Foam Crete is easily sawed, cut, carved, and nailed. It offers excellent insulation values rivaling polyurethane and polystyrene board insulations. It currently costs thirty to fifty cents per pound or $3 per cubic foot.

Zeller International
Main Street, Box Z, Downsville, NY 13755; (607) 363-7792

SYNDECRETE

Syndecrete is a lightweight material developed by David Hertz, president of Syndesis Studios in Santa Monica, California. Its principal constituents are fly ash and polypropylene fiber. A variety of recycled materials such as metal shavings, plastic trimmings, glass chips, and wood chips can be added to create custom products that have been described as modern variations on Italian terrazzo. Up to 40 percent recycled material is used. Pigments can be added for color variations.

Almost anything that can be cast can be made in Syndecrete: countertops, furniture, tiles, tubs, showers, sinks, planters. Syndecrete is strong enough to be cast as non-load-bearing walls or nonstructural panels.

Syndecrete is more resistant to potential chipping and cracking than conventional concrete, tile, or stone, and has a workability more akin to wood than concrete. According to the manufacturer, Syndecrete is chemically inert and does not emit health-threatening indoor air pollutants.

The cost of Syndecrete varies according to the particular product selection; floor tiles range from $10 to $25 per square foot, and a three-foot countertop/sink combination runs approximately $250 to $350.

Syndesis Studios
2908 Colorado Avenue, Santa Monica, CA 90404-3616; (213) 829-9932

WOOD CHIP AND CEMENT BLOCKS

Faswall is one wall system consisting of blocks made of wood chips and portland cement. The blocks are stacked and filled with conventional concrete. In load-bearing walls, reinforcing steel is placed in the cores of the stacked blocks before the concrete is added. Non-load-bearing blocks are filled with insulation. The technology was developed in Switzerland and has been used in Europe for more than 80 years. It is now being manufactured in two plants in the United States: one in Georgia and the other in Iowa.

Faswall can be used above or below grade.

When used above ground, it achieves insulation values as high as R-39; below ground, R-17. It is one quarter the weight of conventional concrete and can be cut with conventional carpentry tools. It is a good soundproofing material. It is noncombustible and insect resistant. It requires no surface coatings, although it is less attractive than AAC and so you probably would want to apply some kind of finish.

Faswall has the same environmental problems of all concrete products, and unlike AAC it requires the additional step of casting in place. Theoretically it is recyclable, but less easily than ACC if the wood chips are to be separated out.

Faswall currently is about twice as expensive as conventional concrete. Costs range from $2 to $3 per square foot.

A similar wood chip and cement block system has been developed by a company called Advanced Concrete Technologies. These blocks are designed primarily for foundation work.

Faswall Concrete Systems
Insulholz Beton International, Inc.
P.O. Box 88, Windsor, SC 29856;
(803) 642-9346

INSTEEL 3D

Insteel 3D is a new breed of reinforced-concrete construction panel with load-bearing capacity. The panels consist of modified, CFC-free, noncombustible expanded polystyrene insulation sandwiched between welded steel wire mesh. The ready-made, lightweight panels are fastened together and a coat of sprayed-on concrete, or shotcrete, is applied. As a bonus, the wire is made of recycled steel.

The standard 4 x 8-inch panel weighs only 38 pounds, or 1.2 pounds per square foot. The insulation value varies according to the thickness of the expanded polystyrene core. For example, a panel with a $2\frac{1}{2}$-inch core and $1\frac{1}{2}$ inches of shotcrete on both sides achieves insulation values of R-11, while one with a 4-inch core and 2 inches of shotcrete on both sides rates R-18. The 3D panel also offers good soundproofing.

Cost savings can be as high as 30 percent over concrete block. Because the panels are lightweight and easy to cut, they can be erected by unskilled laborers easily and quickly. For these reasons, Habitat for Humanity, the international nonprofit housing organization, uses the panel system extensively. What's more, while comparably priced homes in south Florida were destroyed by Hurricane Andrew, those constructed with 3D panels remained intact.

Because Insteel 3D panels are factory assembled, there is less damage by cement trucks and other heavy vehicles at the building site. Because it is lightweight, it is easily transported to the site. Scraps cut from the panels can be used where needed, much as small pieces of gypsum wallboard are.

There are a few environmental minuses involved with Insteel 3D. For example, recycling the five-layer walls would be complicated by the fact that the materials would have to be separated. What's more, there is no data on whether electromagnetic fields pose health problems in homes enveloped in steel-wire mesh.

Insteel Construction Systems, Inc.
41890 Enterprise Circle, S., S-210, Temecula, CA 92590; (909) 676-7876; or 2610 Sidney Lanier Drive, Brunswick, GA 31525; (912) 264-3772

INSTAPERM

Like Insteel 3D, Instaperm is an assembly construction system using a cement-based formula and steel wire mesh. But unlike the former, Instaperm is fabricated on site and utilizes foamed concrete (the Instaperm), along with a number of proprietary products, including Instainsulation, a lightweight foam insulation; Instawire, the galvanized mesh; and Instawrap, an air infiltration barrier. The system was developed by chemist Robert Fondiller of Instaperm in New York City as an affordable construction alternative.

This technology requires a specially equipped truck that carries the mesh-forming equipment, the foaming equipment, the sprayers, and so on. It is intended to be an all-in-one construction system to minimize the number of products and manufacturers and tradespeople, slash installation time, and avoid the need for heavy equipment at the work site. Its environmental appeal is related to this all-in-one approach that minimizes site damage and enables 90 percent of the building, from walls to driveway, to be produced efficiently and quickly from a single "package."

Like Foam Crete, Instaperm is partially comprised of synthetic polymers that the manufacturer claims are nontoxic. Instaperm, like Insteel 3D, raises concerns regarding recyclability and electromagnetic fields.

Instaperm, Inc.
200 West 58th Street, New York, NY 10019; (212) 586-6650

All American Concrete Corporation
19449 Glenwood–Chicago Heights Road, Glenwood, IL 60425; (800) 640-0422, or (708) 756-0555

ALTERNATIVES TO CONCRETE
Buildings made of earth have a simple, basic, instinctive appeal. Earthen construction is particularly appropriate for dry climates. In areas where humidity and precipitation are more prevalent, mold and mildew can be a problem if the walls are not extremely well waterproofed and treated with mildewcides.

ADOBE
Adobe is a traditional building material in the hot and arid climate zone, which extends from southern California through the heart of the Desert Southwest through central Texas. In most of this area, it is not only dry but there are also large temperature swings from day to night. Adobe construction takes advantage of the wide daily differences in temperature. The earthen block walls are so massive that the sun can shine on them all day long without appreciably heating up the inside of the house. After dark, some of the solar heat keeps the interior warm, and the rest is drained right back outside by the cold night sky and cool desert air. By morning, the walls are ready to absorb another day's worth of intense sunshine. Adobe and other earth-based houses that rely substantially on this natural principle of heating and cooling are once again being constructed throughout the Desert Southwest.

Adobe houses are the ultimate in resource efficiency; the adobe blocks can be made entirely of renewable materials found on site — soil, straw, and other plant fibers. The earthen mixture is set in forms and left to cure under the sun. The cured bricks are then laid one upon the other to form walls and usually finished with a protective coat of stucco on the outside and plaster on the inside. Adobe blocks are rot-, vermin-, fire-, and soundproof. And recycling obviously poses no problem: they can be reused easily, or left to return to the earth.

A system called Terrablock was developed by Foam Crete inventor Gary Zeller to reduce the time and labor involved in traditional adobe making. A single piece of portable equipment, the Terrablock Duplex Machine, can produce six to ten blocks per minute. The hopper holds enough earth for ten minutes of continuous operation and can be filled by hand with a shovel or with a front-end-loading tractor. A heavy-duty screen filters out rocks and other debris and pours the mixture into molds. The resulting blocks, which measure approximately 4 x 6 x 12 inches, are lifted by specially designed tongs from the machine's conveyors. They can be used right away or stacked for later use. The blocks can be dry-stacked or laid with traditional mortar. The walls can be finished with plaster or stucco.

Rammed Earth
265 West 18th St. #3, Tucson, AZ 85701; (602) 623-2784

Southwest Solaradobe School
P.O. Box 153, Bosque, NM 87006; (505) 252-1382

Terrablock, Zeller International
Main Street, P.O. Box Z, Downsville, NY 13755; (607) 363-7792

RAMMED EARTH

Rammed earth refers to a number of similar technologies developed to reduce drastically the amount of time and labor required by traditional adobe construction. What these technologies have in common is the formation of earthen blocks or complete walls using extreme pressure or compression.

Rammed earth techniques developed by David Easton of the Terra Group in Napa, California, dispense with two of the primary drawbacks of conventional adobe block — its unsuitability for use in multi-story houses and for homes in earthquake zones. The walls are steel reinforced, and cement is added to the soil for extra strength before it is compacted by pneumatic tampers in removable plywood forms.

Michael Reynolds of Solar Survival Architecture in Taos, New Mexico, has developed a rammed earth technique that makes use of automobile tires from disposal sites. The tires are stacked and filled with earth, forming a wall even thicker and wider than normal with excellent thermal mass and insulating properties. The tire walls are finished on the outside with stucco and on the inside with plaster. The tires, however, are a potential source of indoor air pollution.

Rammed earth structures share all of the advantages of traditional adobe and can be produced a lot more cheaply and quickly.

Environmental Design and Construction
601 23rd Street, Fairfield, IA 52556; (515) 472-8217

Rammed Earth
265 West 18th Street #3, Tucson, AZ 85701; (602) 623-2784

Solar Survival Architecture
P.O. Box 1041, Taos, NM 87571; (505) 758-9870

The Terra Group
1058 2nd Avenue, Napa, CA 94558; (707) 224-2532

STRAW/CLAY BLOCK

Bales of hay coated with clay plaster for protection have been used in building for centuries in Europe and Asia. In an updated technique developed by George Swanson of Environmental Design and Construction, the hay bales are highly compressed and treated with fire retardant, mildewcide, fungicide, and insecticide. The clay is applied as thickly as necessary for thermal mass and good insulation value. Walls made of straw and clay are also extremely soundproof. The protective chemicals and poisons, however, make this building product less attractive.

Environmental Design and Construction
601 23rd Street, Fairfield, IA 52556; (515) 472-8217

WOOD PRODUCTS

When the first Europeans arrived in North America, they marveled over the vast forests that covered much of the continent, full of ancient trees nine feet in diameter or more. The log cabin became the house of American myth: chopped from the forest primeval, built without sawn timbers or nails, it is a symbol of the strong character and independence of the pioneers. The typical house, whether gingerbread cottage or rustic bungalow, is still built with wood, inside and out. Wood is used for beams, joists, studs, and rafters; for sheathing,

SORTING OUT WOODS

All lumber is divided into two categories: softwood and hardwood. Hardwoods are usually denser and harder than softwoods.

SOFTWOODS

Softwood comes from conifers, trees that have thin, needle-shaped leaves and produce their seeds in cones. Conifers are generally evergreen; that is, they usually bear leaves all year long.

More than 75 percent of the wood used for building — and virtually all wood used for framing and other rough construction — is softwood. In the western United States, the structural timber of choice is Douglas fir. In the last magnificent old-growth stands of the Pacific Northwest, this tree grows to well over two hundred feet high with girths of six feet or more. Back East, various species of pine native to the southeastern states are widely used for rough construction.

A variety of softwoods are also employed for finish products such as moldings, doors, and cabinets. A few extremely rot-resistant softwoods, especially redwood, western red cedar, and cypress, traditionally have been used for siding, roofing, decking, and other exterior applications.

HARDWOODS

Hardwoods are broad-leaved. In the temperate forests of the United States, they are also deciduous, meaning they lose their leaves in the autumn. Hardwoods account for about 25 percent of total U.S. lumber production, mostly trees native to the eastern states, especially oak and walnut. Many hardwoods have a much more attractive grain pattern than softwoods and are used for decorative purposes such as moldings and the outside veneers of doors and paneling. Because they are also harder, they are employed as well for floorboards, stair treads, and the like, which must take a lot of abuse.

A number of the most widely used softwood and hardwood species come from tropical rainforests and old-growth forests in the Pacific Northwest. Others, such as cypress and butternut, are species whose populations have declined or are seriously threatened in the deciduous forests of the eastern United States (see "Woods of Concern," page 101).

roof shingles, clapboards, and other siding; for interior furnishings from floorboards to wall paneling to cabinetry to tables; and for outdoor furnishings, including decks, playsets, and fences.

And yet the wooden dwelling's days are numbered. As the ancient forests of the Pacific Northwest have shrunk, so has the supply of giant, old-growth trees that have been the mainstay of the building industry. Current timber harvest rates don't allow enough time for trees to produce the dense heartwood that is prized for structural timber. As a result, builders are forced to rely increasingly on lower-quality trees from commercial forests, on less desirable species once left to rot on the forest floor after the most valuable trees were harvested, and on building products manufactured from what were once lumber mill wastes. Indeed, one way that historians will date the houses of this era, increasingly constructed of composites, laminates, and other manufactured wood products — or other materials entirely — is by distinguishing them from the solid-wood structures of the past.

COMPOSITE WOOD PRODUCTS

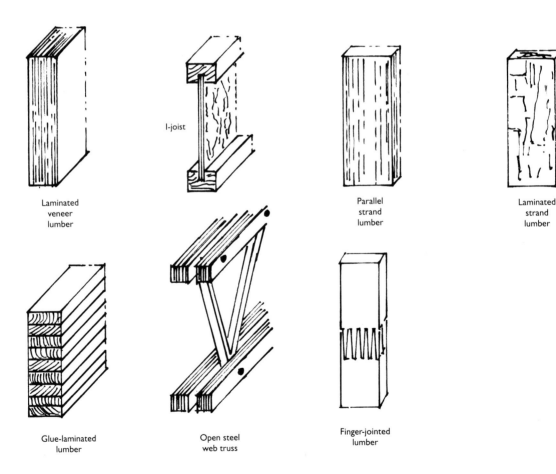

Laminated
veneer
lumber

I-joist

Parallel
strand
lumber

Laminated
strand
lumber

Glue-laminated
lumber

Open steel
web truss

Finger-jointed
lumber

Conventional Products

Solid wood is still used throughout the American house. For three or four decades, plywood has been the sheathing material of choice. Plywood, along with other manufactured wood panels, is assuming ever more important roles in residential construction. This section focuses on the various home products made from solid wood, their ecological ramifications and alternatives. For information on plywood, particleboard, and other manufactured wood panels, which present their own environmental problems, see pages 120–125.

Environmental Concerns

Wood would seem to be the perfect material for the environmental home. It's handsome. It's constantly renewed by nature. It links a home with the surrounding landscape, especially in forested areas, enhancing the occupants' connection to the natural world. If responsible forestry practices have been employed in producing the timber and the particular tree species used are in plentiful supply, wood *is* environmentally preferable to competing structural materials like steel, aluminum, and conventional concrete. But often it is not.

Conventional Forestry Practices

Traditionally, forests both here and abroad have been managed for maximum timber yield (and profit). In order to be truly a renewable resource, wood must be managed on a sustainable yield basis — that is, it should not be cut down faster than it can be regrown or replaced by nature in the wild. The term "sustainable yield" has only recently entered the lexicon of forestry schools, and it's just beginning to filter down into the ranks of foresters in the field. Most forest lands are still managed primarily for just one species and one size of tree in what are called even-aged stands, drastically limiting the diversity of wildflowers and wildlife that can live there. Ecologically managed forests include many species, as well as trees of all ages. They also include a number of other environmental protections, such as riparian zones off-limits to logging to prevent erosion and damage to salmon and trout habitat, and zones of critical habitat for rare and endangered species.

Threatened Old-Growth Temperate Forests

In 1600, about half, or 1.1 billion acres, of what is now the United States was covered with forests. Today, approximately one third, or 730 million acres, of the country is forested. Yet only a tiny fraction of what is left is so-called ancient or old-growth forest — forest that has never been farmed or cut and includes a multi-layered canopy with different aged trees, including many giants hundreds of years old. One reason these forests are important ecologically is because they are so biologically diverse, supporting vast numbers of plant and animal species.

Timber production in the Pacific Northwest has received a great deal of attention in recent years because the region encompasses the nation's last remaining old-growth forest outside of Alaska. Towering redwoods and Douglas firs, two of the most important timber species in the country and some of the oldest trees in the world, live in these forests, as well as the spotted owl, anadromous salmon, and other endangered or declining species. The Pacific Northwest is also one region of the country where timber harvesting has exceeded timber growth in recent decades. Only about 13 percent of the ancient forests in Oregon and Washington remain — some 2.5 million acres, and shrinking.

Threatened Tropical Forests

According to the National Academy of Sciences, each year at least 50 million acres of rainforest — an area the size of Nevada — disappears. If the destruction proceeds at the

WOODS TO LOOK FOR

TROPICAL

Many rainforest specialists believe that it is possible to take the pressure off of overexploited tropical hardwoods and reduce the destruction of tropical forests by using lesser-known species. Anything that adds to the value of these forests, the reasoning goes, strengthens the case for conserving them. By buying these woods, and thereby creating a market for them, architects, interior designers, homeowners, and other wood users can increase the value of tropical forests in general and perhaps prevent them from being burned to the ground once the most valuable trees have been cut. The following are some of the less familiar woods now available in the United States:

Billy Webb (Sweetia panamensis)

The heartwood is olive-brown, somewhat waxy looking, and usually rather sharply demarcated from the yellowish sapwood. Suitable for sills, flooring, and general construction. Highly resistant to termites and durable in contact with the ground.

Breadnut (Brosimum spp.)

Both the heartwood and sapwood are a uniform yellowish white. Good for flooring, furniture and cabinetry, and veneers.

Cabbage Bark (Lonchocarpus spp.)

The heartwood has handsome yellowish brown to dark reddish-brown stripes. Flooring and furniture are among its suggested uses.

Chakte Koc (Sickingia salvadorensis)

A moderately heavy wood with a fine texture. The color is an attractive deep red with wavy light and dark brown streaks. Similar in color to padauk.

Chechem (Metopium brownei)

Chechem heartwood is a deep variegated brown, red, or golden brown with black lines of varying width. The texture is fine and uniform, with barely visible pores. Good for flooring, furniture, and fancy veneers.

Chontaquiro (Diplotropis martiusii)

A moderately heavy hardwood with medium-coarse to coarse texture. Light to dark brown in color and very resistant to rot. An attractive alternative to mahogany or teak for outdoors.

Dao (Dracontolomelon Puberulum)

This wood is a rich gray-brown, accented by heavy black streaks and pinkish overtones. Sometimes referred to as pal dao or New Guinea walnut, it is moderately dense and durable and very stable.

Dillenia (Dillenia spp.)

A fine-textured red to purplish wood with a striking ribboned or fiddleback grain. Unfinished wood has an interesting pinkish salmon color. Also known as simpoh, it resembles rosewood.

Granadillo (Platymiscium spp.)

This lustrous wood is bright red to reddish or purplish brown, more or less distinctly striped. Suitable for fine furniture and cabinet work and decorative veneers.

Kamarere (Eucalyptus deglupta)

This red-brown wood resembles mahogany in appearance and comes from one of the

WOODS OF CONCERN

While architects and homeowners clamor for lists of endangered wood species to avoid, many experts are hesitant to compile them. Trees become threatened when they are not being harvested sustainably. Within the same country, a tree species may be endangered in one area and responsibly harvested in another. In the case of domestic trees like Douglas fir, it is not the species itself but rather the old-growth forests from which it is cut that are endangered. Use the following list as a rough guide. The important thing is to buy the woods listed below from certified or reputable sources.

TROPICAL

Afrormosia	(*Pericopsis elata*)
Andiroba	(*Carapa guianensis*)
Ebony	(*Diospyros* spp.)
Iroko	(*Chlorophora excelsa, C. regia*)
Lauan	(*Shorea* spp.)
Lignum Vitae	(*Guaiacum officinale*)
Macawood	(*Platymiscium* spp.)
Mahogany	(*Khaya* spp.)
Mahogany	(*Swietenia humilis*)
Okoume	(*Aucoumea klaineana*)
Padauk	(*Pterocarpus soyauxii* and other *Pterocarpus* species)
Piquia	(*Caryocar costaricensis*)
Ramin	(*Gonstylus* spp.)
Rosewood	(*Dalbergia stegvensonii, D. nigra, D. latifolia*)
Sapele	(*Entandrophragma cylindricum*)
Teak	(*Tectona grandis*)
Utile	(*Entandrophragma utile*)

TEMPERATE

Butternut	(*Juglans cinerea*)
Cypress	(*Taxodium distichum*)
Douglas fir	(*Pseudotsuga menziesii*)
Holly	(*Ilex opaca*)
Port Orford cedar	(*Chamaecyparis lawsoniana*)
Redwood	(*Sequoia sempervirens*)
Western red cedar	(*Thuja plicata*)

world's fastest-growing trees. When quartersawn, it has a striking ribbon grain figure. Kamarere is a very abundant tree in New Guinea, and lends itself to a wide range of uses.

■ **Kwila** (*Intsia bijuga*)

Also known as merbau, this wood has golden to dark brown heartwood. The grain is often wavy. Extremely hard and durable and often used for flooring, furniture, and other applications in which strength and stability are of key importance.

■ **Malas** (*Homalium foetidum*)

This wood is orange- to red-brown. Like kwila, malas is very dense and durable. It is ideal for flooring and, because it takes preservatives well, ideal for decking, exterior millwork, and outdoor furniture.

■ **Mersawa** (*Anisoptera* spp.)

Mersawa is a heavy cream-colored wood with even grain. It has occasional rose streaks and darkens over time. Similar to the better-known tropical wood ramin.

■ Monkey Pod *(Pithecellobium saman)*

This wood is golden to dark brown in color with wavy grain. Monkey pod has long been a favorite wood among turners, but it is also suitable for a wide range of uses, including fine furniture, cabinetry, and interior trim.

■ Narra *(Pterocarpus indicus)*

A prized member of the *Padauk* genus. Also known as New Guinea rosewood and narravitali, the wood varies from golden brown to rich red-brown, with a sharply defined cream-colored sapwood. A beautiful wood, it is as durable as teak. The highly figured burls are prized in Europe.

■ New Guinea Red Cedar *(Cedrela toona)*

A pinkish-red wood with some streaking and an even grain. Red cedar is a hardwood, unrelated to our native coniferous western red cedar. Quite durable but relatively weak, easy to saw and work and lighter in weight than many softwoods — almost like balsa. Used widely abroad for fine furniture, cabinet work, and interior trim.

■ Pencil Cedar *(Palaquium* spp.)

Pencil cedar is the common name of several species in the *Palaquium* genus. It is also known as nyatoh. The wood, pale pinkish brown, is well suited for moldings and joinery.

■ Planchonia *(Planchonia* spp.)

An attractive, heavy timber with a rich, classic mahogany look.

■ Santa Maria *(Calophyllum brasiliense)*

The heartwood varies in color from pink or yellowish pink to brick red or rich reddish brown. Widely used in the tropics for general construction, flooring, and furniture.

■ Shihuahuaco *(Dipteryx* spp.)

A moderately heavy hardwood with medium-coarse to coarse texture. Yellowish to golden brown in color and very rot resistant.

■ Taun *(Pometia pinnata)*

This attractive, light-colored wood varies from pale pinkish to deep red-brown. Highly valued in Japan for furniture, it is equally suitable for millwork and cabinets. Taun is the most abundant tree in Papua New Guinea and an excellent substitute for mahogany.

■ Terminalia *(Terminalia brassii* and other species)

A light brown wood suitable for indoor trim and paneling.

■ Water Gum *(Syszium* spp.)

A pale to grayish-brown wood with occasional red and purple streaks. An excellent wood for fine furniture and other custom uses.

■ Wild Tambran *(Pithecellobium arboreum)*

The heartwood is dark reddish brown sometimes with alternating lighter reddish brown to golden brown. Has a luster resembling Honduras mahogany. Uses include fine furniture and cabinetry, interior trim, paneling, and fancy veneers.

TEMPERATE

In native North American forests, as in the tropical forests, it makes ecological sense to seek out woods that are usually passed over in the quest for big-money species. By buying these underutilized woods, you help increase the value of every acre. This in turn will help justify good husbandry and make it less likely that loggers will simply move on to new areas once a few cash trees have been harvested. Buying less familiar domestic woods can open up new markets for logging communities whose economies have been ravaged by the overharvesting of commercially prized species. Using domestic species can also take some of the pressure off of overexploited tropical woods. The following native woods are both commercially available and suitable for flooring, cabinetry, and other kinds of architectural millwork.

American Beech (*Fagus grandifolia*)

This is a light-colored wood with no pronounced grain, suitable for formal interiors in which highly grained woods are too rustic looking. It is hard, but more sensitive to moisture than teak or cherry, so don't use it for windows, in the bathroom, or outdoors, where it will get wet. American beech is a good substitute for mahogany.

Black Oak (*Quercus kelloggii*)

This native Californian and Oregonian oak is a good substitute for red oak from the eastern forests, which it very much resembles.

California Bay (*Umbellularia californica*)

This tree is known by a variety of other names, including California laurel, California olive, myrtle, Oregon myrtle, and pepperwood. It has a strong pungent odor when bruised. The heartwood is yellow-brown.

Hickory (*Carya* spp.)

A handful of the several dozen hickory species that grow in North America are commercially available, including pecan hickory (*Carya illinoinensis*), pignut hickory (*C. glabra*), bitternut hickory (*C. cordiformis*), and shagbark hickory (*C. ovata*). They range in color from white to chestnut. Hickory is extremely hard and tough but like beech should not get wet. It is a good substitute for ash.

Madrone (*Arbutus menziesii*)

The heartwood of madrone is light to medium reddish brown and resembles cherry. It is extremely easy to work. Madrone is in plentiful supply in Pacific Northwest forests.

Oregon Maple (*Acer macrophyllum*)

Also called canyon or big-leaf maple, this West Coast species is a good substitute for the eastern hard or rock maple (the sugar maple, *Acer saccharum*). Like eastern red maple, another good substitute, it is not as hard as rock maple.

Red Elm (*Ulmus rubra*)

Red elm is a pretty, reddish chestnut wood with a pronounced grain pattern. It is as stable as, and a good substitute for, red oak. It is also an excellent substitute for mahogany and native hard or rock elm. It does require more sanding as finishes are likely to raise the grain.

Red Maple (*Acer rubrum*)

Red maple is a good substitute for rock or hard maple. Although it is softer, it is less affected by moisture. In addition, it tends to have a more curly grain figure. Also known as swamp or soft maple, red maple is usually light tan to warm red but has a broad color range, from almost paper white to blues and greens as well as reds. Like beech, it is not appropriate for moist areas.

Tan Oak (*Lithocarpus densiflorus*)

Like madrone, this tree is plentiful in forests of the Pacific Northwest. The sapwood is light reddish brown and turns darker with age; the heartwood is also darker. Tan oak is very easy to work.

White Oak (*Quercus garryana*)

This western oak native to California is denser than eastern red oak and more rot resistant.

ROT-RESISTANT WOODS

Tropical

Billy Webb	(*Sweetia panamensis*)
Chontaquiro	(*Diplotropis martiusii*)
Narra	(*Pterocarpus indicus*)
Malas	(*Homalium foetidum*)
Shihuahuaco	(*Dipteryx* spp.)

Domestic

Aromatic cedar	(*Juniperus virginiana*)
Black locust	(*Robinia pseudoacacia*)
Character or industrial grade cherry	
	(*Prunus serotina*)
Osage orange	(*Maclura pomifera*)
Sassafras	(*Sassafras albidum*)

current rate of over 150 acres per minute, these forests will be gone before the middle of the next century.

Tropical hardwoods are especially prized for their coloring and beautiful grain. The typical American home may well contain a lengthy list of products made at least in part of timber harvested from the forests of Borneo or Amazonia: veneer on the dining table, desk, stereo cabinet, bedroom set, and wall paneling; less costly items such as salad bowls and cheese boards; and major components of the structure itself, including cabinetry, decking, and some plywoods.

In 1986, the total estimated value of tropical hardwood products imported to the United States was approximately $2.4 billion. Of that, furniture accounted for about one billion; plywood, veneers and joinery almost another billion; and raw logs and lumber about $430 million. These figures do not include still another group of imports: products manufactured in non-tropical countries but made of tropical woods.

The tropical forests comprise the rainforests, crowded with coconut and date palms and hundreds of other trees, including mahogany and other species vital to the timber industry. These lush forests receive at least 80 inches of rainfall a year, and their trees flower and fruit continuously. Some trees soar to more than 200 feet, and the vegetation is so dense that usually less than one percent of the sunlight that strikes the upper canopy filters down to the forest floor. Rainforests occur in a wide band around the equator, particularly the Amazon lowlands in South America, the Congo basin in Africa, and the Malay archipelago between Asia and Australia. The latter two areas have been heavily depleted.

Another tropical ecosystem, the monsoon forest, is the native habitat of such prized woods as teak. These forests receive somewhat less precipitation than the rainforests, and there is an annual dry season of three months or more when most trees shed their leaves. The canopy of the monsoon forest is lower, and there is more light penetration. The sunlight reaching the forest floor promotes a dense, junglelike tangle of undergrowth. Monsoon forests were once extensive in southeast Asia and to a lesser extent other tropical regions.

It is estimated that more than half of the species on earth can be found in tropical forests, many still to be discovered and studied by scientists. Yet according to the United Nations Food and Agriculture Organization (FAO), these forests occupied only 6.7 percent, 3.9 million square miles, of the Earth's land masses in 1980. One hectare (about 2.5 acres) of old-growth Brazilian rainforest may contain more than 200 tree species alone, compared to 10 or 15 species per hectare in our own temperate forests.

About 1,000 different tribes depend on tropical forests worldwide for survival. Tropical forests yield not only wood but also a wealth of other raw materials, among them essential oils, latexes, waxes and dyes, and vegetables, fruits, nuts, and spices. About one quarter of all modern medicines are based on natural chemicals from tropical forests, and scientists expect to find many, many more as only about one percent of the known plant species have been chemically screened for medicinal value.

The ancient forests of the Pacific Northwest and the tropics supply us with much of the lumber with which we build and decorate our homes. How we manage — or mismanage — them can have great consequences for world biodiversity and the survival of individual plant and animal species, as well as other environmental problems such as global warming. For example, scientists suspect that tropical forests act as a vast "sink" for carbon dioxide that would otherwise remain in the earth's atmosphere and contribute to the so-called greenhouse effect. Depletion of ancient forests around the world would also have devastating effects on the people — including the loggers in our own Pacific Northwest — who depend on them for their livelihood or basic subsistence.

Environmental Products

Whether you're looking for a two-by-four or a strip of molding, a few rules of thumb can help you avoid making the kinds of choices that contribute to the destruction of ancient forests and threatened species. For details, see the individual product categories that follow.

- Use wood efficiently. Buy only as much as you need, and ask your architect or builder to design for standard lumber lengths so that there will be as few cut-offs as possible. There's an additional benefit to reducing the amount of wood you use: you'll offset at least some of the recent lumber price increases and save money.

- Be especially wary about buying redwood and other tropical and temperate species of concern listed in "Woods of Concern," page 101.

- Look for alternative composite wood products made of laminated woods, lower-grade wood fibers, and lumber mill scraps. These products make use of wood that might otherwise be discarded. Examples are finger-jointed studs and I-joists.

- Look, too, for lumber salvaged from demolished buildings or from loose boards stockpiled in old barn lofts, mills, lumberyards, and furniture shops. Across the country, buildings constructed in the nineteenth century and earlier are being razed. Many are in a state of partial collapse, particularly abandoned barns. Factory buildings, especially in the Northeast, are being vacated as industry moves south or abroad. In the South, tobacco barns and cotton warehouses are coming down. An array of species, hardwood and softwood alike, are salvaged from these sites.

   Salvaged woods often have a distinct advantage: they are the highest-quality representatives of their species, having originated from old-growth forests many years ago, when, for example, redwoods and Douglas firs twelve to fifteen feet in diameter or more were common in the Pacific Northwest. The heartwood of these old trees is unsurpassed.

Alas, it is not always easy to find salvaged wood, as there is no national or international effort of any size to track down sources and inventory supplies. Sales of salvaged lumber occur mostly through small networks of buyers and sellers. When a buyer is looking for a particular species of wood, with specific dimensions, grade, and character, potential supplies are tracked down by the modern-day equivalents of word-of-mouth: telephone, fax, and computer modem. As a result, it can take time to find what you need, and when you do, you can expect to pay more for these woods than you would for conventional lumber.

It's also important to understand that salvaged wood often shows blemishes, bolt holes, traces of paint, and other discolorations. To most admirers of precious woods, these are not defects; on the contrary, they add character and patina. Salvaged wood being used as structural members should be oversized to pass muster with local building inspectors, many of whom are unfamiliar with salvaged sources and their characteristics. In fact, salvaged wood typically is stronger than its conventional counterparts. It's usually very dry and well seasoned — it may be so tight-grained and hard, in fact, that pre-drilling will be necessary for fasteners.

- Wood felled by hurricanes, tornados, earthquakes, volcanoes, and other natural disasters, which typically is burned or carted off as waste to a landfill, is yet another environmentally sound choice. The volumes of this so-called waste can be enormous; Hurricane Hugo alone, which lashed the East Coast, particularly South Carolina, in 1990, took down more than 200,000 board feet of trees.

Consider, too, wood from trees cleared for utility lines, razed by developers for building and landscaping, or harvested by the few remaining horse loggers. Wood harvested by the latter is considered "naturally felled" because these old-time loggers operate with horses and oxen, not trucks and skidders, and therefore do minimal damage to forest ecosystems.

Together, these sources have the potential to become a major supply of logs for lumber. For now, however, like salvaged wood, naturally felled lumber can be difficult to find and, if you're lucky enough to locate it by phone or fax, expensive. Another drawback is that the wood is green. Unless it has been stockpiled at a mill and either air dried or kiln dried with the regular inventory, you'll need to make arrangements to have it dried. Despite the drawbacks, if you're looking for large-dimensioned logs for either structural or aesthetic reasons, or if you simply like the idea of rescuing good wood from an ignominious end at the local landfill, don't rule out naturally felled woods. They certainly make great conversation pieces.

- Use lumber and other products made from woods that have been deemed as sustainably produced by a reputable certification organization. Right now, relatively little of such lumber is available, but supplies are growing every day.

Just what is a sustainably managed wood source? Here in the United States, the Tropical Forest Foundation, a group that includes industry groups and professional organizations such as the American Institute of Architects (AIA), are working to nail down specific criteria for sustainable forestry management. For the purposes of the report on tropical woods in its *Environmental Resource Guide*, the AIA considers a wood source to have been sustainably managed only when (1) management techniques that foster the long-

LOG HOMES

There aren't too many Americans who haven't dreamed at one time or another about living in a log cabin. Log cabins are romantic, handsome, and rustic. They make you feel like you are living close to the land. However, log cabins are often made of woods in increasingly short supply.

The whole logs used to construct a cabin are often cypress, Western red cedar, fir, and aspen. Using a rot-resistant wood such as cedar or cypress is ideal, but this can add to pressure on remaining old-growth forests in the Pacific Northwest or dwindling cypress stands in the Southeast. Logs that are more vulnerable to the weather are treated with toxic wood preservatives. The mortar used as chinking contains synthetic compounds that can make the air inside the home unhealthy.

If you are building a log home, use a wood like aspen, which grows relatively quickly and is in plentiful supply. Treat the logs with a less toxic wood preservative (see "Wood Preservatives," pages 133–134). And use a mortar that does not emit health-threatening pollutants. An entire line of safe products for log homes is available from Perma-Chink Systems, Inc. If you live in the West, write the Western Division at 17635 N.E. 67th Court, Redmond, WA 98052, or call (800) 548-1231. In the East, write the Eastern Division, 1605 Prosser Road, Knoxville, TN 37914, or call (800) 548-3554.

term health of the forest, such as rotations, are used; (2) harvesting methods minimize damage to surrounding trees and maximize the forest's ability to regenerate; (3) the affected area is either reforested or prepared for natural regeneration after logging; (4) efforts are made to harvest and market not just the most valuable species but also a variety of alternative species; (5) wood processing operations are in place near the harvested area to enhance job opportunities and the local economy; (6) wood waste from processing is shared with the local population for use as fuel; and (7) harvesting and production of non-timber forest products by the local population is encouraged. Similar criteria are used by the two major organizations currently engaged in certifying sources of tropical woods: the Rainforest Alliance and Scientific Certification Systems.

- Consider lesser-known species of tropical and domestic woods. These are often good substitutes for threatened or overexploited species. In addition, if loggers are encouraged to harvest a range of trees, they will be less likely to move into new areas of forest once the few commercially valuable trees have come down. Alternative tropical and domestic woods are just beginning to be discovered by architects, designers, and homeowners (see "Woods to Look For," pages 100–104).

- Don't forget about woods native to your own area. In many parts of the country, locally harvested and sawn lumber is readily available from small milling operations — all you have to do is look. Using these woods can reduce the pressure on sensitive forests, and slashes the amount of energy wasted in getting remote timber supplies to your home. It also boosts your local economy.

- Finally, request "character grades" rather than "clear veneer grades." The latter are largely free of knots and other "defects" and are consistent in color. Yet as Paul Fuge of Plaza Hardwoods points out, these attributes are not normal in a tree; diversity in color is natural. As a result of our insistence on clear veneer grades, lumber companies go after the infrequent tree that is likely to have high-grade lumber, and the rest of the forest can be damaged or destroyed in the rush to get to it. Character grades usually have some blemishes, but as the name suggests, they give the wood character. Character grade woods are normally relegated to such uses as pallets and railroad ties, but there's no reason they can't do justice to flooring, cabinetry, and furniture.

Framing Lumber

Framing accounts for as much as 70 percent of the approximately 11,000 board feet of lumber used to build a typical house (and the average house size is increasing). Multiply that by the 90 percent of new homes in the United States that are wood framed and you're talking about a lot of wood. The traditional practice has been to mill this large-dimension lumber from old-growth trees such as Douglas fir. Because the supply of these magnificent specimens has dwindled rapidly and the workability and structural integrity of the quickly grown wood from commercial forests that is being sold in its place pales in comparison, the industry is turning to products that use wood fiber a lot more efficiently, products made of smaller-diameter second- and third-growth wood, scraps, and other sources. These new materials often perform better than the old-growth lumber they've

replaced. They do generally cost more than conventional lumber. Yet as the price of dimensional lumber continues to increase, the environmental alternatives are becoming more and more economical.

I-joists — These substitutes for joists and roof rafters get their name from their shape; a cross-section view of the joists looks like a capital letter "I." The top and bottom of the I are made of solid wood or laminated veneer lumber; the vertical part, of plywood or oriented strand board. I-joists can carry the same load as — or more than — comparable solid-sawn floor joists and roof rafters, using about 50 percent less wood fiber.

Manufactured trusses — Trusses are made of top and bottom boards (usually two-by-fours) joined together by boards (generally also two-by-fours) or tubular steel in a sawtooth configuration. They are good substitutes for roof rafters and floor joists.

Laminated veneer lumber (also known as LVLs and parallams) and **glue-laminated lumber** (Glu-lams) — Both these composite wood products are manufactured by gluing together small pieces to create large structural lumber. Larger pieces of timber are required for glue-laminated lumber than for laminated veneer lumber, but it is still a step in the right direction. Like I-joists, these engineered products can span longer distances and support heavier loads than their solid-sawn counterparts. They're an excellent alternative to solid timbers for beams and columns.

Finger-jointed lumber — These are composed of short pieces of lumber that once would have been consigned to the trash heap. The pieces have been joined together to create standard-sized two-by-four and two-by-six studs. Finger-jointed studs are as strong as solid lumber of the same grades, and they are resistant to stress, warp, and twist.

Steel studs and framing — These are still another option, but only if they are made of recycled steel (see "Recycled Steel," pages 135–137).

Roofing and Siding

From rambling, shingle-style houses on the Northeast coast bleached silvery gray by the sun and salt spray to redwood contemporaries on the California coast, wood is the time-honored exterior finish for the American home. Because they resist cupping and checking and are impervious to rot, the best woods for siding and roofing are western red cedar and redwood. Yet supplies of good quality wood of both species are dependent on surviving old-growth forests in the Pacific Northwest and western Canada.

For this reason, and because wood poses a fire hazard when used on the roof, it's a good idea to avoid wood roofing. The environmental alternatives are slate, salvaged slate, terra-cotta tile, recycled aluminum, concrete tiles, and wood fiber–cement composite slates. Lightweight aluminum shingles are available with up to 100 percent recycled material. Fiber-cement composite slates are long lasting and come with warranties of up to sixty years, and much of the wood fiber comes from sawmill wastes.

From the environmental point of view you'd do best to avoid wood siding as well. Instead, consider natural stucco, stone, especially salvaged stone, brick, or lightweight concrete (see "Concrete," pages 90–96). If you can't do without clapboard, shingles, or board-and-batten siding, look for a certified or salvaged wood source, or a domestic wood in plentiful supply. If you decide to go this route, keep in mind that many woods must be painted or treated with preservatives to make them less vulnerable to the weather. Use one of the less toxic paints or preservatives now on the market (see "Enamel Paints," pages 185–187, and "Wood Preservatives," pages 133–134).

Flooring

From the fanciest inlaid parquets in southern plantation houses to the simplest wide planks in New England saltboxes, wood has been the traditional material for American floors. From the environmental standpoint, wood is still one of the best flooring materials, especially if you use a certified wood source and finish it with one of the new, less-toxic products (see "Wood Floor Finishes," page 193). Several of the sources recommended in the supplier's list that follows offer salvaged woods, including heart pine and teak, suitable for flooring, but this supply is far from inexhaustible. Look for character grades rather than clear grades of prized domestic hardwoods such as white oak, rock maple, and cherry. A number of less familiar domestic hardwoods also make beautiful floors, especially red maple, red elm, beech, and hickory (see "Woods to Look For," pages 100–104).

Wood Trim

Some of the wood used in the typical house serves as trim or decoration. Moldings placed around doors and windows, crown moldings (where the walls meet the ceiling) and baseboard moldings (where the walls meet the floor), chair rails, wainscot, and fireplace mantels fall into this category.

Use wood sparingly for these purposes, and to full effect. The best decorative woods, from the environmental point of view, are the less familiar domestic hardwoods, such as red maple and red elm. There are also a number of lesser-known tropical woods that make beautiful substitutes for mahogany and other threatened tropical species (see "Woods to Look For," pages 100–104). Finger-jointed molding and trim is another resource-conserving alternative.

Furniture and Cabinetry

Many of the exquisite woods that traditionally have been used for moldings and other interior finishes have also been fashioned into furniture and cabinetry: tropical woods like mahogany, rosewood, and ebony; domestic woods such as cherry, maple, and oak. Teak and redwood have been the preferred species for outdoor furniture. Some of these tropical species are already or may become endangered in all or part of their range due to commercial overexploitation. Many of the favored domestic hardwoods have also been heavily depleted.

Whether you're constructing your own furniture or shopping for it, look for pieces made with handsome but lesser-known species from both tropical and temperate forests. A number of companies that produce furniture using wood from sustainable sources have been certified by the Rainforest Alliance (see "Furniture," pages 210–212).

Decking

The wood deck has become one of the most popular amenities of the modern American home. Decks and other outdoor "rooms," from patios and terraces to verandas and gazebos, provide beautiful living space and enable us to live closer to nature. However, populations of the rot-resistant woods best for decks, as well as other exterior uses, such as cypress and redwood, have been heavily depleted. Other, less impervious woods are often employed for decking, but to last, these must be treated with wood preservatives, which have their own environmental problems (see "Wood Preservatives," pages 133–134).

So what are the environmental choices? For starters, consider alternatives to the wooden deck — brick or stone patios or terraces, for example. Stone lasts a lot longer than wood and doesn't require toxic finishes. If you must have a wooden deck, make every effort to locate a source of salvaged redwood or cypress. Another alternative is to select lesser-known tropical and domestic species that are durable, rot resistant, and insect resistant and do not require surface coatings (see "Rot-Resistant Woods," page

104). Consider, too, one of the new recycled-plastic lumber products (see "Recycled-Plastic Lumber," pages 137–140).

Pressure-Treated Woods

Pressure-treated wood is the dimensional lumber in lumberyards that has a greenish cast. Used widely outdoors not only for decks but also playsets, fences, and furniture, it has been treated with poisonous preservatives to make it less vulnerable to insects and rot. If at all possible, avoid pressure-treated wood when building your arbor or garden bench. For a more detailed discussion and a list of suppliers of alternative products, see "Wood Preservatives," pages 133–134.

Suppliers

COMPOSITE WOODS

TJM Trus Joist Macmillan
9777 West Chinden Boulevard, P.O. Box 60, Boise, ID 83707; (208) 375-4450 or (800) 628-3997
Micro=Lam laminated veneer lumber for headers and beams, Parallam parallel strand lumber for headers, beams, columns, and posts, Silent Floor truss joists, and Timberstrand laminated strand lumber for rim board and headers.

Alpine Structures, Inc.
317 Providence Road, Oxford, NC 27565; (800) 672-2326
I-joists for roof and floor framing, as well as headers, Timbermax laminated veneer lumber, which can be used for a variety of structural purposes.

Boise Cascade Corporation
P.O. Box 2400, White City, OR 97503; (800) 232-0788

I-joists and Versa-Lam laminated veneer lumber for headers, ridge, floor, and roof beams, roof rafters, and floor joists.

Champion International Corporation
P.O. Box 1593, Tacoma, WA 98401; (206) 572-8300
A complete line of finger-jointed lumber.

Georgia Pacific
133 Peachtree Street NE, P.O. Box 105605, Atlanta, GA 30348-5605; (800) 447-2882
GP-Lam laminated veneer lumber beams and headers, and Wood I Beam I-joists.

Jager Industries, Inc.
8835 MacLeod Trail SW, Calgary, Alberta, Canada T2H 0M3; (403) 259-0700
TTS Wood I-joists and Super Joists for floor and roof joists.

Louisiana-Pacific
2706 Highway 421 North, Wilmington, NC 28401; (800) 999-9105
Gang-Lam laminated veneer lumber and GNI Joists for floors and roofs.

Nascor, Inc.
1212 34th Avenue SW, Calgary, Alberta, Canada T2G 1V7; (403) 243-8919
A line of I-joists for use as floor joists and roof rafters.

Standard Structures, Inc.
P.O. Box K, Santa Rosa, CA 95402; (707) 544-2982
A variety of manufactured wood products, including SSI Joists and finger-jointed and laminated framing lumber.

Unadilla Laminated Products
P.O. Box K, 32 Plifton Street, Unadilla, NY 13849; (607) 369-9341
Glued laminated columns, trusses, arches, purlins, and beams.

Willamette Industries, Inc.
P.O. Box 277, Saginaw, OR 97472; (503) 744-4655
Glue-laminated beams, columns, and headers.

CERTIFIED TROPICAL WOODS

Scientific Certification Systems, formerly Green Cross, one of the nation's leading independent evaluators of products claimed to be environmental by their manufacturers, has certified one forestry operation in Mexico, Plan Piloto Forestal. (SCS has also certified four domestic forestry operations, listed under "Domestic Woods," pages 116–118), as well as a number of manufacturers and retailers who sell products made from wood harvested from these certified forests. The latter companies are listed under "Domestic Woods," "Environmental Flooring," page 119, "Miscellaneous Wood Products," pages 119–120, and "Furniture," pages 210–212. Manufacturers or suppliers must pay a fee to be evaluated by SCS. This can restrict some environmental companies, often startup enterprises with limited capital. For more information on how the SCS certification process works, or on the evaluations of specific products, write to Scientific Certification Systems, The Ordway Building, One Kaiser Plaza, Suite 901, Oakland, CA 94612, or call (510) 832-1415.

Plan Piloto Forestal Estatal
Infiernillo 157, Esqu. Efrain Aguilar, Chetumal, Quintana Roo, Mexico; 011-52-983-24424
Woods include Honduran mahogany, Spanish cedar and, many lesser-known species.

The Rainforest Alliance certifies sources of Smart Wood and companies that sell Smart

Wood products — that is, products made exclusively of certified woods. To date, six forestry operations have been certified. For a list of exclusive Smart Wood companies, see "Furniture," page 00. For more information on certification and approved products, contact the Rainforest Alliance at 65 Bleecker Street, New York, NY 10012; (212) 677-1900. The forestry operations currently certified that offer lumber products are:

State Forestry Corporation, Java, Indonesia
Contact Lynn-Nusantara Marketing Co., Inc., 21 East 28th Avenue, Suite D, Eugene, OR 97405; (503) 686-9886
Plantation-grown teak, mahogany, rosewood, and pine available in semi-finished or finished products.

Plan Forestal Estatal, Quintana Roo, Mexico.
See the listing under Scientific Certification Systems, above.

Masurina
Papua New Guinea. Contact Eco-timber, 350 Treat Street, San Francisco, CA 94110; (415) 864-4900
Lesser-known species available in lumber.

Proyecto Desarrollo Bosque Latifoliado
(Broadleaf Forest Development Project)
P.O. Box 427, La Ceiba, Honduras; (504) 43-1032.
Lesser-known species available in lumber.

Tropical American Tree Farms
Campo Real and Santo Domingo, Costa Rica
Contact Tropical American Tree Farms, 717 City Park Avenue, Columbus, OH 43206; (614) 443-5300
Plantations in early growth stages, so wood may not be available yet. Trees planted include teak, Brazilian rosewood, and many lesser-known species.

AMACOL, Ltd.
Portel, Para Brazil
Tropical woods available in veneer (for corestock ply) and plywood. See listing under "Plywood, Particleboard, and Other Manufactured Wood Panels," page 123.

The Rainforest Alliance also certifies companies that sell Smart Wood products as well as non-certified tropical wood products:

Northern Hardwood Lumber, Inc.
520 Matthew Street, Santa Clara, CA 95050; (408) 727-2211
Wholesaler and retailer of lumber from lesser-known species.

Sea Star Trading Company
P.O. Box 513, Newport, OR 97365; (503) 265-9616
Wholesaler and retailer of over a dozen lesser-known species in various forms, including lumber, veneer, and various finished products.

Eco-Timber
350 Treat Street, San Francisco, CA 94110; (415) 864-4900
Wholesaler and retailer of lumber in lesser-known species.

Wise Wood
P.O. Box 1271, McHenry, IL 60050; (815) 344-4943
Wholesaler and retailer of lumber in lesser-known species.

The San Francisco–based Rainforest Action Network (RAN) and the Vermont-based Woodworkers Alliance for Rainforest Protection (WARP) do not certify tropical woods but do distribute lists of reputable suppliers. Some of them sell certified timber as well as wood that is not from ecologically managed forests, so be sure to inquire about the source of the woods you are interested in. The suppliers include:

A & M Wood Specialty
P.O. Box 32040 Cambridge, Ontario, Canada N3H 5M2; (519) 653-9322
Woods from Papua New Guinea.

Almquist Lumber
100 Taylor Way, Blue Lake, CA 95525; (707) 668-5652
Wood from Plan Piloto in Mexico.

Appalachian Interiors
201 Amy Drive, Maryville, TN 37801; (615) 984-4989
Woods from Peru.

Berea Hardwoods
6367 Eastland Road, Brook Park, OH 44142; (216) 234-7949
Wood from Plan Piloto in Mexico.

Crosscut Hardwoods
3065 NW Front Avenue, Portland, OR 97210; (503) 224-9663
Woods from Mexico.

Cut & Dried Hardwoods
241 South Cedros, Solana Beach, CA 92075; (619) 481-0442
Wood from Plan Piloto in Mexico.

Demerara Timber Ltd.
Lot 1, Water Street and Battery Road, Kingston, Georgetown, Guyana; 011-592-2-53835
Greenheart, purpleheart, demerara mahogany, and lesser-known species.

Ecological Trading Company
659 Newark Road, Lincoln LN6 8SA, England; 011-44-0522-501-850
Woods from Papua New Guinea, Mexico, Peru, and Honduras.

Edensaw Woods
211 Seton Road, Port Townsend, WA 98368; (800) 950-3336 or (206) 385-7878
Wood from small-scale community operation in Peru and Plan Piloto in Mexico.

EnviResource
110 Madison Avenue North, Bainbridge Island, WA 98110; (206) 842-9785
Perpetua tropical woods.

Gilmer Wood
2211 NW Saint Helens, Portland, OR 97210; (503) 274-1271
Six species of wood from small-scale community operation in Peru.

Handloggers Hardwood Lumber
135 E. Sir Francis Drake Boulevard, Larkspur, CA 94939; (415) 461-1180
Wood from small-scale operations in Belize and Mexico. Excellent selections for flooring.

Karp Woodworks
229 Lowland Street, Holliston, MA 01746; (508) 429-8636
Wood from Mexico and Papua New Guinea.

Pittsford Lumber
500 State Street, Pittsford, NY 14534; (716) 381-3489
Woods from Mexico, Peru, and Mozambique.

Sound Products Ltd.
1800 C Egg Lake Road, Friday Harbor, WA 98250; (206) 378-2693
Perpetua hardwoods.

Tree Products Hardwoods
P.O. Box 772, Eugene, OR 97440; (503) 689-8515
Perpetua hardwoods.

Wildwoods Co.
445 I Street, Arcata, CA 95521; (707) 822-9541
Woods from Mexico.

Wilson Woodworks
18 Hydeville Road, P.O. Box 273, Stafford, CT 06075; (800) 545-1861 or (203) 684-9112
Wood from Plan Piloto in Mexico.

Woodcastle Forest Products
34030 Excor Road, Highway 34, Albany, OR 97321; (503) 926-5488
Perpetua hardwoods.

Woodworkers Source
5402 South 40th Street, Phoenix, AZ 85040; (800) 423-2450 or (602) 437-4415.
Wood from small-scale community operation in Peru and Plan Piloto in Mexico.

Rio Rivuma is a Boston-based conservation group that offers environmentally harvested woods from the eastern Transvaal and Mozambique. Logs, air- and kiln-dried lumber, dressed stock, and veneer are available. For more information, contact Rio Rivuma at 229 A Street #2C, Boston, MA 02210-1722, or call (617) 451-2549.

SALVAGED WOODS

It's estimated that there are 30 to 50 suppliers across the United States. Environmental Construction Outfitters is the first national organization to begin establishing a network of these suppliers and their inventory. For information, call (800) 238-5008 or (212) 334-9659.

Michael Evenson
Box 191, Redway, CA 95560; (707) 923-2979
Salvaged redwood and Douglas fir from structures in northern California, most built during the 1950s from trees that grew in the Klamath River watershed, just north of Redwood National Park. The salvaged Douglas fir is known for its structural strength, and many pieces are stamped "Durable Fir #1," a grade that no longer exists as there are too few remaining trees like these to harvest and mill. Also available is Alaskan yellow cedar, an excellent wood for cabinetry, and redwood planks salvaged from wine tanks.

Plaza Hardwood, Inc.
5 Enebro Court, Santa Fe, NM 87505; (800) 662-6306 or (505) 989-7885
Specialize in salvaging character woods that would otherwise be used as pulpwood for paper or for pallets, including white oak, hickory, cherry, sapwood teak, and ash. Also available is reclaimed antique lumber, including heart pine, chestnut, white pine, and oak.

Aged Woods, Inc.
2331 E. Market Street, York, PA 17402; (800) 233-9307 or (717) 840-0330
Mostly barnwood for flooring. Pines, oak, chestnut, poplar, cherry, hickory, walnut.

Bronx 2000
1809 Carter Avenue, Bronx, NY 10457; (718) 731-3931
Big City Forest Products made with wood reclaimed from discarded shipping pallets and crates. Mahogany and other tropical hardwoods, oak, poplar, pine, and fir are combined into solid butcher block, giving each piece a distinctive look.

Byers and Son
P.O. Box 449, Trinidad, CA 95570; (707) 822-9007
Douglas fir.

Caldwell Building Wreckers
195 Bayshore Boulevard, San Francisco, CA 94124; (415) 550-6777
Douglas fir, maple, redwood.

J. Chapman, Inc.
P.O. Box 173, Ware Shoals, SC 29692; (803)
456-3492
Long-leaf yellow pine.

Details
1350 Elm Street, Napa, CA 94559; (707) 226-
9443
Salvaged woods.

Duluth Timber Company
P.O. Box 16717, Duluth, MN 55816; (218)
727-2145
Primarily Douglas fir, some heart pine, redwood, cedar, cypress, and various hardwoods.

Florida Ridge
4114 Bridges Road, Groveland, FL 34736;
(904) 787-4251
Antique heart pine, cypress.

Goodwin Lumber Co.
Route 2, Box 119, Micanopy, FL 32667; (800)
336-3118 or (904) 466-0339
Antique heart pine, antique heart cypress from
logs cut 80 years ago.

Into the Woods
300 North Water Street, Petaluma, CA 94952;
(707) 763-0159
Salvaged urban wood.

Jefferson Lumber Company
1500 W. Mott Road, Mt. Shasta, CA 96067;
(916) 235-0609
Timbers, beams, and lumber milled from wood,
primarily Douglas fir and ponderosa pine, reclaimed from old buildings in the Pacific Northwest.

The Joinery Co.
P.O. Box 518, Tarboro, NC 27886; (800) 726-
7463 or (919) 823-3306
Flooring, railing, trim from 200-year-old heart
pine.

Karp Woodworks
229 Lowland Street, Holliston, MA 01746;
(508) 429-8636
Salvaged urban wood.

S. Mack
Chasehill Farm, Ashaway, RI 02804; (401)
377-2331
Pine, oak, chestnut.

Mountain Lumber
P.O. Box 289, Ruckersville, VA 22968; (804)
985-3646
Antique heart pine, oak, chestnut flooring.

G. R. Plume Co.
Architectural Timbers Division, 1301 Meador
Avenue, Suite B11 & 12, Bellingham, WA
98226; (206) 676-5658
Salvaged woods.

Recycle the Barn People
P.O. Box 294, St. Peter's, PA 19470; (215)
286-5600
Salvaged woods.

Recycled Lumber Works
596 Park Boulevard, Ukiah, CA 95482; (707)
462-2567

Sourcebank
1325 Imola Avenue West #109, Napa, CA
94559; (707) 226-9582
Serves as broker for companies that salvage.
Wood species vary monthly.

Urban Forest Woodworks
585 West 3900 South #6, Murray, UT 84123;
(801) 266-5650
Salvaged woods.

Urban Ore
1333 6th Street, Berkeley, CA 94710; (510)
559-4460
Douglas fir, redwood.

Vintage Lumber Co.
9507 Woodsboro Road, Fredrick, MD 21701;
(800) 499-7859 or (301) 898-7859
Salvaged woods.

Wesco Used Lumber
911 Ohio Avenue, Richmond, CA 94530;
(510) 235-9995
Douglas fir, redwood.

NATURALLY FELLED LUMBER
The best places to look for naturally felled woods are utility companies, local land developers, and the National Park Service, as well as state and local parks. Environmental Construction Outfitters is the first national organization to begin a nationwide network of these sources and their inventory. Call ECO at (800) 238-5008 or (212) 334-9659.

DOMESTIC WOODS
Four domestic forestry operations have been certified by Scientific Certification Systems:

Collins Pine
Contact Wade Mosby, 1618 SW 1st Avenue, Suite 300, Portland, OR 97201; (503) 227-1219
Woods include sugar pine, ponderosa pine, white fir, incense cedar, Douglas fir, and red fir.

Kane Hardwood
Contact Wade Mosby, 1618 SW 1st Avenue, Suite 300, Portland, OR 97201; (503) 227-1219
Woods include cherry, red and white oak, ash, hard and soft maple, poplar, basswood, cucumber, eastern white pine, and eastern hemlock.

Menominee Tribal Enterprises
Contact Roger Waukau, P.O. Box 10, Keshena, WI 54150; (715) 756-2314, ext. 142

Woods include cherry, butternut, cedar, hard and soft maple, red oak, birch, beech, basswood, hemlock, white pine, ash, and aspen.

Pingree Family Ownership
Contact John Cashwell, President, or John McNulty, Vice President, Seven Islands Land Management Company, 304 Hancock Street, Suite 2A, P.O. Box 1168, Bangor, ME 04402-1168; (207) 947-0541
Woods include black, white, and red spruce, yellow and paper birch, balsam fir, eastern white pine, eastern hemlock, northern white cedar, sugar and red maple, American beech, brown and white ash, apsen, and northern red oak.

The following companies sell wood from forests certified by Scientific Certification Systems:

Gilles Begin Ltd.
Contact Jean Collin, C.P./P.O. Box 100 Clair Noveau, New Brunswick, Canada E0L JB0; (506) 992-2113
Hardwood lumber.

The Home Depot (selected stores)
Contact Mark Eisen, One Paces West, 2727 Paces Ferry Road NW, Atlanta, GA 30339-4053; (404) 801-5871
Pine shelving and white fir dimensional lumber.

Les Produits Forestiers Becesco
Contact Michel Poulin, 51 Rue de Moulin, Saint-Juste, Quebec, Canada, G0R 3H0
Hardwood lumber.

Materiaux Blanchet, Inc.
Contact Daniel Michaud, 1030 Elgin SUD, C.P. 430, St. Pamphile, Quebec, Canada, G0R 3X0; (418) 356-3344
Spruce and balsam fir lumber.

Rumney Wood Products
Contact Lawrence Berndt, P.O. Box 211,
Harrington Road, Cornish, NH 03746;
(603) 675-6206
Hardwood and softwood lumber.

Stripling Blake Lumber
Contact Don Strickland, 3400 Steck Avenue,
Austin, TX 78758; (512) 465-4200
White fir dimensional lumber.

A handful of organizations are developing standards for sustainable domestic forestry operations. For more information, write:

Institute for Sustainable Forestry
P.O. Box 1580, Redway, CA 95560; (707)
923-4719

Rogue Institute for Ecology and Economy
P.O. Box 3213, Ashland, OR 97520; (503)
482-6031

SGS Silvaconsult, Ltd.
Magdalen Centre, Oxford Science Park, Oxford OX4 4GA, England; 011-44-865-202-345

Silva Forest Foundation
RR1, Winlaw, British Columbia, Canada V0G
2J0; (604) 226-7770

The Soil Association
86 Colston Street, Bristol B51 5BB, England;
011-44-117-929-0661

Other recommended suppliers of domestic
woods:

Allan Branscomb
597 West 29th Avenue, Eugene, OR 97405;
(503) 687-1422
Douglas fir and white oak lumber certified by
the Rogue Institute.

Dartington Home Woods
Contact Silvanus, 15 Link House, Leat Street,
Tiverton, Devon EX16 5LG England; 011-44-
0884-257-344
Douglas fir, larch, western red cedar, redwood,
and various hardwoods certified by The Soil Association.

Michael Evenson
P.O. Box 191, Redway, CA 95560; (707) 923-
2979
Second-growth redwood and Douglas fir.

Forest Trust Wood Products Brokerage
P.O. Box 519, Santa Fe, NM 87504; (505)
983-8992 or (505) 983-3111
Southwest woods from community operations
in New Mexico.

Warren Fuller
11750 Hillcrest Road, Medford, OR 97404;
(503) 772-8577
Second-growth wood certified by the Rogue Institute.

Into the Woods
300 N. Water Street, Petaluma, CA 94952;
(707) 763-0159
Flooring in madrone, western maple, tan oak.

Keweenaw Land Assoc.
East 5090 Jackson Road, Ironwood, MI
49938; (906) 932-3410
Northern hardwoods and softwoods.

Plaza Hardwood, Inc.
5 Enebro Court, Santa Fe, NM 87505; (800)
662-6306 or (505) 989-7885
A variety of less familiar domestic species suitable for flooring, including red elm, hickory, red
maple, and beech.

Tosten Brothers
P.O. Box 156, Miranda, CA 95553; (707) 943-3093
Native West Coast woods.

ENVIRONMENTAL ROOFING
American Cem-Wood Products
3615 Pacific Boulevard SW, Albany, OR 97321; (800) 367-3471 or (503) 928-6397
Cemwood Shakes and Permatek shakes made of roughly two thirds portland cement and one third wood fiber derived from waste sawmill chips.

Classic Products
P.O. Box 701, Piqua, OH 45356; (800) 543-8938 or (513) 773-9840
Aluminum roofing made from recycled beverage cans. Designed and colored to look like wood shakes.

Eternit, Inc.
Excelsior Industrial Park, P.O. Box 679, Blandon, PA 19510; (800) 233-3155 or (215) 926-0100
Fiber-cement composite roofing slates made from portland cement and wood fiber. Also available as corrugated roofing panels.

Everest Roofing
P.O. Box 2265, Irwindale, PA 91706; (800) 767-0267
Cal-Shake cedar-style roof tile.

FibreCem Corporation
11000-I South Commerce Boulevard, P.O. Box 411368, Charlotte, NC; (800) 346-6147 or (704) 527-2727
Fiber-cement composite roofing slates composed of portland cement, wood fiber, silica, and other additives.

James Hardie Building Products
10901 Elm Avenue, Montana, CA 92335; (800) 426-4051 or (714) 582-0731
Hardishake fiber-cement composite slates that look like wood shakes.

Masonite Corporation
1 South Wacker Drive, Chicago, IL 60606; (800) 255-0785
Woodruf Traditional Roofing Shingles made from wood fibers bonded together under heat and pressure to create a shingle that's denser and more durable than natural wood.

Maxitile, Inc.
17141 Kings View Avenue, Carson, CA 90746; (800) 338-8453; or 4153 L B McLeod Road, Orlando, FL 32811; (800) 845-6293
Maxitile composite roofing consisting of portland cement, silica, and wood fiber designed to look like two-piece terra-cotta mission tile.

Metal Sales Manufacturing Corporation
Denver, CO; (800) 289-7663
Steel metal roofing panel designed to look like clay tiles.

New England Slate Co.
Burr Pond Road, Sudbury, VT 05733; (802) 247-8809
Slate roofing.

Re-Con Building Products
P.O. Box 1094, Sumas, WA 98295; (800) 347-3373
Fiber-cement roofing that looks like slate.

Scandinavian Profiling Systems
1951 Hamburg Turnpike, Buffalo, NY 14218; (716) 826-2593
Galvanized steel tile look-alike.

Supradur
P.O. Box 908, Rye, NY 10580; (800) 223-1948 or (914) 967-8230
Fiber-cement composite roofing shingles available in slate, shake, and other traditional styles.

Vermont Structural Slate Co., Inc.
P.O. Box 98, Prospect Street, Fair Haven, VT 05743; (800) 343-1900 or (802) 265-4933
Slate roofing.

ENVIRONMENTAL FLOORING

Aged Woods
2331 E. Market Street, York, PA 17402; (800) 233-9307 or (717) 840-0330

J. Chapman, Inc.
P.O. Box 173, Ware Shoals, SC 29692; (803) 456-3492

Coastal Millworks
1335 Marietta Boulevard NW, Atlanta, GA 30318; (404) 351-8400

Green River Lumber
P.O. Box 329, Locust Hill Road, Great Barrington, MA 01230; (413) 528-9000

Hendrickson Naturlich
8031 Mill Station Road, Sebastopol, CA 95472; (707) 829-3959

Into the Woods
300 N. Water Street, Petaluma, CA 94952; (707) 763-0159

The Joinery Co.
P.O. Box 518, Tarboro, NC 27886; (800) 726-7463

Mountain Lumber
P.O. Box 289, Ruckersville, VA 22968; (804) 985-3646

Plaza Hardwood
5 Enebro Court, Santa Fe, NM 87505; (800) 662-6306 or (505) 466-7885

A.E. Sampson & Son, Ltd.
P.O. Box 1010, Warren, ME 04804; (207) 273-4000

Sound Products Ltd.
1800 C Egg Lake Road, Friday Harbor, WA 98250; (206) 378-2693

Superior Flooring
803 Jefferson Street, Wausau, WI 54401; (800) 247-4705

Vintage Lumber Co.
9507 Woodsboro Road, Fredrick, MD 21701; (800) 499-7859 or (301) 898-7859

Y.W.I.
P.O. Box 211, Harrington Road, Cornish, NH 03746; (603) 675-6206

See also companies listed under "Certified Tropical Woods," "Salvaged Woods," and "Domestic Woods."

MISCELLANEOUS WOOD PRODUCTS

Doors

Chindwell Doors, Johor, Malaysia
Contact B & Q Plc., Portswood House, One Hampshire Corporate Park, Chandlers Ford, Eastleigh, Hampshire SO5 3YX, England; 011-44-0703-256-256
Rubberwood doors.

Royal Mahogany Products, Inc.
Contact Al Barrencchea, 6145-1 Northbelt Parkway, Norcross, GA 30017; (404) 729-1600

Doors made from tropical woods certified by Scientific Certification Systems.

Molding and Millwork

Bainings Community-Based Ecoforestry Project
Rabaul, Papua New Guinea. Contact B & Q Plc., Portswood House, One Hampshire Corporate Park, Chandlers Ford, Eastleigh, Hampshire SO5 3YX, England; 011-44-0703-256-256

Chinquapin Mountain Designs
13401 Highway 66, Ashland, OR 97520; (503) 482-6220

Colonial Craft
2772 Fairview Avenue North, St. Paul, MN 55113; (612) 631-3110

Warren Fullmer
11750 Hillcrest Road, Medford OR 97504; (503) 772-8577

A. E. Sampson & Son, Ltd.
Contact Paul Sampson, P.O. Box 1010, Warren, ME 04804; (207) 273-4000

Wood Shingles

Industries Maibec, Inc.
Contact Charles Tardif, Rue de l'Eglise Est, St.-Pamphile, Cte. l'Islet, Quebec, Canada G0R 3X0; (418) 356-3331
Eastern white cedar shingles from wood certified by Scientific Certification Systems.

PLYWOOD, PARTICLEBOARD, AND OTHER MANUFACTURED WOOD PANELS

Before the advent of plywood and other manufactured wood products, houses were not only framed but also sheathed with solid wood. The transition in the 1930s and forties from 1x6s and 1x8s installed on the diagonal for structural strength to plywood for sheathing was a sign that the large logs needed for sawn lumber were already becoming scarce. What's more, only about half of each log milled can be sawed into boards, meaning that as much as half of each log was being wasted. Technological advances in the past fifty or so years, as well as the ever-decreasing supply of large-diameter logs, have led the timber industry to develop products that make use of by-products from sawmills and get the most out of every tree.

During the early days of plywood production in this country, only huge specimens of Douglas fir from old-growth forests in the Pacific Northwest were used. By the mid-1960s, the search for new woods brought the plywood industry to the Southeast to take advantage of abundant southern pines. As even the smaller logs needed for plywood production have become harder to come by, new products comprised of wood chips and flakes, other plant fibers, and even waste papers have become available. In all of these resource-conserving, and generally cheaper, alternative materials, however, the wood fibers are bonded together with petroleum-based, synthetic resins, presenting potential health and environmental problems.

Conventional Products

Plywood
Plywood is made of thin veneers of wood glued together in a sandwich. The grain of each layer runs perpendicular to the grain of its neighbors,

giving this material great structural strength.

There are two main types of plywood: hardwood and softwood. In general, softwood plywoods are used for structural applications, whereas hardwood versions are used for decorative purposes, especially cabinetry and paneling. Softwood plywood is now the standard material for wall, floor, and roof sheathing. It's also used for countertops, shelving, and exterior siding. In 1988, 706 million cubic feet of softwood plywood was produced, all derived from indigenous North American trees.

Virtually all softwood plywood is manufactured with a phenolic adhesive, which consists of phenol formaldehyde resins plus wheat flour, ground bark, and other fillers. Because phenol formaldehyde resin is waterproof and extremely durable, it is used both indoors and outdoors.

Wall Paneling

Hardwood plywood faced with a veneer of beautifully grained wood is often used to finish interior walls. It is also used in the construction of cabinetry and furniture. Seventy percent of all hardwood plywood is composed of a veneer core, often lauan, a tropical wood also known as meranti. Fourteen percent has a particleboard core and another fourteen percent a core of medium density fiberboard. The inner core is faced with a veneer of tropical wood such as mahogany and teak, or domestic species including cherry or oak.

In 1988, American consumption of hardwood plywood amounted to 78 million cubic feet, of which less than half, or 29 million cubic feet, was made of native North American trees. Virtually all hardwood plywood is manufactured with urea-formaldehyde resin adhesives.

Particleboard

Particleboard is composed of wood chips or other wood residues bound together by an adhesive. Extremely flat, dense, and stable when dry, it is used primarily indoors. About three quarters of all particleboard is used as core stock for furniture and cabinets, and 8 percent is used as an underlayment for floors.

Although particleboard makes good use of wood fiber that might otherwise be discarded as waste, its major drawback is that a relatively high 6 to 9 percent of the weight of the finished product comprises the adhesive that glues it all together — almost always potentially health-threatening urea-formaldehyde resin.

Waferboard

Also known as flakeboard, waferboard is made by bonding waferlike wood chips, or flakes, together with a phenol formaldehyde resin. The flakes range in size and thickness and can be either randomly or directionally oriented.

Waferboard is used chiefly as a less expensive substitute for sheathing grades of plywood. Because it has two smooth sides it is also used as shelving.

Oriented-Strand Board

Oriented-strand board (OSB) is manufactured from compressed, strandlike wood particles bonded with phenol formaldehyde resin. Because OSB is composed of oriented layers of flakes, much like the cross-laminations of plywood, it is exceptionally strong and stable. And because it is substantially cheaper, oriented-strand board has made serious inroads into subflooring and wall and roof sheathing markets once dominated by plywood. The high-speed flaking machines that make the strands for OSB can accept logs as small as four inches in diameter from trees that once had little commercial value. Aspen is the species most commonly used in the North, while southern plants use thinnings from commercial tree farms.

Fiberboard

Fiberboard is a close relative of particleboard. It is made from very fine softwood fibers bonded with resin. The wood fibers come from culled timber, logging residues, and mill wastes. Other fibers, including agricultural wastes such as sugar cane, as well as waste paper can also be used. Chemical agents are added for strength, moisture resistance, and fire prevention.

There are two major classifications of fiberboard: compressed intermediate or medium density, better known as MDO, and hardboard. MDO is generally used to sheath walls and as subflooring. The most common binder for medium-density fiberboard and hardboard is urea-formaldehyde resin. Hardboard has a hard, flat, temperated finish. It is often used in cabinets and as drawer bottoms. When perforated with evenly spaced round holes, it's known as "pegboard." Fiberboard can be finished with a variety of surface textures, including stucco, or prepainted at the factory for use as exterior siding.

Stress-Skin Structural Panels

These panels are used in "kit," "timber-frame," and other factory-built homes. They were developed to help simplify and speed up construction. The wall panels are usually built at the factory, then shipped to the site for installation onto a structural skeleton.

There are a number of different kinds of stress-skin panel systems. Some consist of exterior sheathing and gypsum wallboard with foam insulation and a vapor barrier sandwiched in between. Another option is a complete wall system with not only exterior siding and sheetrock but also windows, wiring, and other systems built in. Some environmental builders advocate these systems because they often require less wood than conventionally framed and sheathed homes and result in less environmental damage at the building site. What's more, quality is virtually assured because the panels are fabricated under ideal factory conditions.

Health and Environmental Concerns

Indoor Air Pollution

A major environmental concern associated with the use of manufactured wood panels is their potentially adverse effect on the air quality inside your home. The problem is the urea-formaldehyde resin used to bond the veneers, chips, flakes, strands, and fibers together. Wall paneling and other hardwood plywood products, particleboard, and fiberboard — all made with urea-formaldehyde adhesives — are particularly problematic. Urea-formaldehyde is considered a probable human cancer causer by the U.S. Environmental Protection Agency. It has also been linked to a host of other health problems, from headaches and dizziness to nausea and rashes. To make matters worse, particleboard and paneling are used inside the house, where they can release formaldehyde for months or even years.

The Hardwood Plywood Manufacturers Association estimates that formaldehyde emissions from hardwood plywood produced today are 70 to 95 percent lower than those from hardwood plywood produced a decade ago. The National Particleboard Association cites similar reductions. One study of various hardwood plywoods showed formaldehyde emissions in the 0.10 to 0.12 parts per million (ppm) range. Research also suggests that emissions decrease over time. In one study of particleboard, for example, formaldehyde emissions were 0.35 ppm one month after manufacture, 0.15 ppm five months after manufacture, and 0.10 ppm seven months after manufacture. The U.S. Department of Housing and Urban Development (HUD) standard for manufactured housing limits formaldehyde concentrations from particleboard to 0.3 ppm and from decorative plywood paneling to 0.2 ppm. However, some people begin to feel irritation at much lower levels. Irritation has been observed at levels as low as 0.01 ppm.

Phenol formaldehyde adhesives, used in soft-wood plywood, waferboard, and oriented-strand board, are more stable and emit much lower amounts of formaldehyde. Panels bonded with phenol formaldehyde adhesives are not regulated by HUD.

Threatened Old-Growth Forests

Plywood plants, like commercial sawmills, rely on a dwindling resource — large logs. A considerable amount of North American plywood still depends on wood logged from the nation's remaining old-growth forests in the Pacific Northwest.

Threatened Tropical Forests

Prized tropical timbers, including mahogany, are used for the veneers that face wall paneling and other decorative hardwood plywoods. Some are considered endangered or overexploited in all or part of their range. Conservationists are also concerned about the overexploitation of the lauan that typically composes the plywood core.

Environmental Products

Manufactured wood panels are preferable to sawn lumber for many parts of your home because they make better use of harvested trees. Even in this category of materials, however, some choices are better than others.

It is wise to avoid all products made with urea-formaldehyde resin — which means virtually all particleboard and wall paneling and other types of hardwood plywood found in the local lumberyard or home center, unless the label states specifically that they have been bonded with less-toxic resins. Another thing to consider if you're shopping for wall paneling is the particular species from which it is made. Avoid those that are faced with veneers made from tropical tree species that are threatened

by commercial overexploitation (see "Woods of Concern," page 101). Look for plywoods made from lesser-known tropical woods that have been certified by environmental watchdog groups like the Rainforest Alliance.

You can also make a difference by opting for the products that have made the most efficient use of wood resources. In this respect, oriented-strand board, say, is preferable to plywood, which depends on decreasing supplies of large logs. Products that incorporate mill wastes, agricultural wastes, or recycled paper are better yet. In the stress-skin structural panel systems currently available, wood is used more sparingly than it is in conventionally framed homes. However, the stress-skin panels have other drawbacks. The foam insulation incorporated in the panels produces toxic fumes when burning and is made with either ozone-depleting CFCs and HCFCs or smog-producing pentane. In addition, the plywood components may contribute to the levels of formaldehyde in your home. The sandwich-like panels would also be difficult to recycle, were recycling systems in place, because the various materials would have to be separated first.

Suppliers

AMACOL, Ltd.
Portel, Para, Brazil. Contact Larson Wood Products, Inc., 31421 Coburg Bottom Loop, Eugene, OR 97401; (503) 343-5229
Plywood core veneers from a tropical forestry operation certified as sustainable by the Rainforest Alliance.

Buchner Panel Manufacturing
1030 Quesada Avenue, San Francisco, CA 94124; (800) 483-6337
Veneer and plywood panels made from temperate woods.

For other suppliers of certified tropical wood veneers, see "Certified Tropical Woods," pages 111–114.

AFM Corporation
24000 W. Highway 7, Suite 201, Shorewood, MN 55331; (612) 474-0809
Stress-skin panels with plastic foam cores that are not blown with ozone-depleting CFCs or HCFCs. Franchised production facilities are located across the country.

AgriBoard Industries
P.O. Box 645, Fairfield, IA 52556; (515) 472-0363
Compressed straw structural panels.

Bellcomb Technologies, Inc.
70 N. 22nd Avenue, Minneapolis, MN 55411; (612) 521-2425
Offers a patented building system using a series of interlocking kraft-paper honeycomb panels composed of 5 percent paper and 95 percent air. The panels can be faced with a variety of materials. They can be used for interior and exterior walls, floors, roofs, beams, and connecting pieces. This construction system can reduce by two thirds the amount of lumber required for construction of a home.

Eternit, Inc.
Excelsior Industrial Park, P.O. Box 679, Blandon, PA 19510; (800) 233-3155 or (610) 926-0100
Manufacturer of Eterboard, a calcium silicate board that is a healthy alternative to plywood for use both indoors and outdoors.

Evanite Fiber Corporation
P.O. Box E, Corvallis, OR 97339; (503) 655-3383
Hardboard made from waste wood such as pallets, crates, construction debris, sawdust, and plywood manufacturing waste. For use in furniture.

Fibrelam
P.O. Box 2002, Doswell, VA 23047; (804) 876-3135
Makers of Energy Brace, comprised of 25 percent recycled paper, with one side faced with foil. Also available in several strength grades.

Homasote Company
P.O. Box 7240, West Trenton, NJ 08628-0240; (800) 257-9491 or (609) 883-3300
Produces a variety of sheathing products made from 100 percent recycled newsprint, including Easy-ply roof decking, Homasote fiberboard for exterior siding, Homasote 4-way floor decking, a structural subfloor that also deadens sound, and several prefinished wall panelings.

Louisiana-Pacific
1000 Woodfield Road, Suite 134, Schaumburg, IL 60173; (800) 365-7672
Makes Inner-Seal oriented-strand board sheathing using wood fiber from small, fast-growing species such as aspen.

Mansion Industries
P.O. Box 2220, City of Industry, CA 91746-2220; (818) 968-9501
Structural panels made of compressed straw.

Meadowood Industries, Inc.
33242 Red Bridge Road, Albany, OR 97321; (503) 259-1303
Meadowood paneling made from varieties of seasoned rye grass selected for color and strength, suitable for wall paneling, ceilings, room dividers, cabinetry, and furniture. This material has some of the resilience and acoustical properties of wood, with insulation, strength, and fire rating greater than many wood products. And it makes use of an agricultural waste product. Natural florals, ferns, and branches can be inlaid during manufacturing for special decorative effects.

Medite Corporation
P.O. Box 4040, Medford, OR 97501; (800) 676-3339, (503) 773-2522 or (503) 779-9596
Manufacturers of Medite II formaldehyde-free, medium-density fiberboard for cabinets, shelving, and countertop underlayments; and Medex, a waterproof material for non-load-bearing exterior and interior uses.

Oregon Strand Board Co.
34363 Lake Creek Drive, Brownsville, OR 97327; (503) 466-5177 or in the West (800) 533-3374
Uses Douglas fir veneers and mill waste in Comply and Sturd-I-Floor, good choices for roof sheathing and subflooring, respectively, because of their high strength.

PanTerre American, Inc.
2700 Wilson Boulevard, Arlington, VA 22201; (703) 247-3140
Compressed straw structural panels.

Sea-Star Trading offers certified tropical veneers on Meadowood for cabinetry, paneling, and other uses. For information, write Sea-Star Trading Co., P.O. Box 513, Newport, OR 97365; or call (503) 265-9616 or (800) 359-7571.

Simplex Products
P.O. Box 10, Adrian, MI 49221; (517) 263-8881
Thermo-ply sheathing with insulating properties, made from 100 percent recycled kraft paper with foil or polyethylene facings. Useful for sheathing where high strength isn't mandatory. Available in several color-coded strength grades.

Weyerhaeuser
4111 W. Four Mile Road, Grayling, MI 49738; (517) 348-2881
Maker of Structurwood, an oriented-strand board comprised of fiber from small, fast-growing trees. Useful for sheathing.

For wood-conserving construction systems that employ lightweight concrete mixtures, see descriptions of Faswall and Insteel 3D, pages 93–94.

INSULATION

The word "insulation" derives from the Latin word *insula,* or island. Insulation transforms a building into a kind of island, isolated from the outside, both in terms of temperature and sound. In the past, builders achieved this with thick walls made of stone, timber, and earth and roofs of sod and soil. When massive walls become too expensive and cumbersome to erect, builders packed walls with straw, wool, coconut fibers, leaves, or any other readily available material.

Today, insulation can still be provided through the use of massive walls and sod roofs. Adobe and rammed earth homes, for example, are still constructed in the Southwest. But more commonly, builders and architects specify lumber-framed walls and roofs and pack them, as well as crawl spaces, foundation walls, and other areas, with one or more layers of manmade material with good insulating value.

The environmental benefits of insulation are enormous. It's estimated that American homes consume 10.7 quads of energy a year. (A quad is 1,000,000,000,000,000, or one quadrillion, British thermal units, or Btus. A Btu is the standard measure for the amount of heat produced from various types of energy.) That translates to more than 1,835,000,000 barrels of oil a year. Many of our most severe environmental problems — including acid rain, oil spills, and radioactive waste disposal, as well as war — are a direct result of energy production and use. Beefing up the insulation in your home helps reduce energy demand, which in turn helps reduce the environmental damage.

Most Americans have already added some insulation to their houses since the oil shocks of the 1970s sent energy prices soaring. But most homes could benefit from still more. The U.S. Department of Energy publishes a list of minimum recommended insulation levels for houses by zip code. Check your library or call the Energy Efficiency and Renewable Energy Clearinghouse (EEREC) at (800) 428-2525 for recommended levels in your area. Some energy experts recommend that you double these minimum recommendations if you want your home to be truly energy efficient. NATAS or a local energy consultant can help you calculate how much energy can be saved by additional insulation, and how quickly the insulation will pay for itself in lower energy bills. Keep in mind that at some point it makes little sense to upgrade insulation levels if your windows aren't energy efficient; precious heat will simply escape through the glass (in warm regions heat from outside will flow through the glass into your rooms and make your air conditioner work harder). See "Windows," pages 141–144.

There are a variety of types of insulation. Some have greater R-value per inch than others. (R-value, or resistance value, is a measure of a material's ability to keep heat from flowing into or out of a room; the higher the R-value, the more energy conserving the insulation.) Generally, the insulations with the highest R-value per inch are the most expensive. Some insulations are more fireproof or moisture-proof than others. Insulation usually comes in the following forms:

- Blankets or batts — either 16 or 24 inches wide, the standard widths of wall, floor, and ceiling cavities in the typical house
- Loose fill — poured into ceiling and wall cavities from bags or blown in by a contractor
- Rigid boards — high in cost as well as R-value and used most frequently on flat roofs, on the outside of walls, in foundation walls, and below concrete slabs
- Sprayed-in-place foams — sprayed into walls and ceilings by trained professionals

Conventional Products

The most common materials used for thermal insulation are fiberglass, mineral wool, foam, and cellulose.

Fiberglass is manufactured from sand, dolomitic limestone, and borax.

Mineral wool is available in two types: rock wool and slag wool. Rock wool is made from diabase and basalt rock. Slag wool is made mostly from slag from iron-ore blast furnaces.

Foams made into rigid boards are composed of polyisocyanurate, expanded polystyrene (also known as beadboard), extruded polystyrene, and phenolic foams. Polyurethane is the most common blown-in-place foam. Each of these synthetic foam insulations is derived from petrochemical feedstocks.

Cellulose insulation is made primarily from recycled paper and cardboard. Because it is highly flammable, a fire retardant must be added.

Health and Environmental Concerns

Ozone Depletion

All of the synthetic foam insulations are manufactured with blowing or expansion agents. Traditionally, these have been chlorofluorocarbons (CFCs), which have been found to deplete protective ozone in the upper atmosphere. Ozone depletion has been linked to skin cancer, cataracts, and, recent studies suggest, depression of the immune system. There are also a host of possible ecological effects, including a decline in the growth of phytoplankton in the oceans. The U.S. Environmental Protection Agency considers ozone depletion one of the most serious environmental problems.

According to the recent Montreal Protocol on Substances That Deplete the Ozone Layer, a set of conventions accepted by the international community, countries around the world are supposed to eliminate the use of CFCs by the year 2000. About 30 percent of the CFCs currently used in the United States are used in plastic foam insulation. Polyisocyanurate, phenolic, and some polyurethane foams are still made with CFCs. Many manufacturers are looking to partially halogenated hydrochlorofluorocarbons (HCFCs), which deplete ozone to a lesser extent, as substitutes for their chemical cousins, the CFCs. Extruded polystyrene is now blown with HCFC-142b, which is approximately one twentieth as damaging to the ozone layer as CFC-12, which it replaced. In the United States, HCFCs are supposed to be phased out by the year 2030. Beadboard or expanded polystyrene is blown with pentane, which doesn't deplete atmospheric ozone but does contribute to ground-level smog.

Indoor Air Pollution

In the late 1970s, some of the first indications of what is now known as the sick building syndrome came from residents of homes insulated with urea-formaldehyde foam. Headaches, irritation of the eyes and the respiratory tract, and dizziness were common complaints. In many of these homes, levels of formaldehyde, a suspected human carcinogen, proved to be extremely high. Urea-formaldehyde foam insulation is now rarely used in new construction. However, some common insulation materials emit small amounts of formaldehyde as well as other volatile organic compounds (VOCs) that are known irritants (see "A Guide to Home Pollutants," pages 245–249). Fiberglass insulation, for example, often is manufactured with formaldehyde-based resin as a binder; the formaldehyde content of the insulation may approach several percent of the product by weight. The fire retardants used in cellulose insulations, as well as inks and other chemicals in the recycled paper from which they are often made, are potential sources of VOC pollution. It is also important to note that the foam insulations emit potentially deadly gases when they burn.

Asbestos, one of the most thoroughly studied human cancer causers, was once used in insulation materials. Its use in insulation products has been voluntarily discontinued in the United States. While it is no longer available, it may still be a problem in older homes (see "An Environmental Renovation," pages 32–41).

One result of asbestos's decline has been greater use of other man-made mineral fibers such as fiberglass and mineral wool. The primary factor linking asbestos to cancer is its long, thin fibers. Fiberglass and mineral wool also have fibers in the troublesome-size range, and there has been a good deal of controversy in recent years over whether they, too, present health hazards. Based on human and animal studies the consensus seems to be that the low fiber levels in the typical house may be a health problem, but probably not. Fiberglass insulations, however, are being scrutinized. Several European countries ban the use of fiberglass, and California has banned blown-in fiberglass.

Oil Depletion

Synthetic foam insulations are derived from petroleum and natural gas, the very fuels that insulation is intended to conserve. There's something paradoxical about the fact that in an attempt to conserve these precious fuels, we produce tons of insulation made from them. According to the Washington, D.C.–based World Resources Institute, we'll be out of cheap, readily available petroleum by the year 2025 if we continue to consume it at current rates.

Environmental Products

Unfortunately, from an environmental point of view there are no perfect insulations, except perhaps the air used between glass panes in energy-efficient windows. All insulations have some performance drawbacks and environmental limitations. The alternative insulations recommended below are the best solutions available today. Some are made from renewable resources, such as cotton; others are petroleum-based and emit relatively few VOCs and are made without ozone-depleting CFCs; others are noncombustible and release no noxious fibers to the air.

It is not uncommon to use a variety of these newer environmental insulations in a typical home-insulating project because each is most appropriate for a specific application. For example, Air Krete is best used inside closed-up walls in older homes, or inside concrete block walls or foundations. Cotton batts are recommended for new walls, floors, and rafters. Radiant/Low E barriers are appropriate in walls and ceilings for added insulation and vapor barrier protection and around duct work and the water heater. Rigid boards are used under slabs, outside foundation walls, on flat roofs, or on the exterior of walls for added R-value. Although many cellulose insulations are made from recycled paper, they are not listed here under suppliers because they can be a substantial source of VOC emissions.

Suppliers

BLANKETS AND BATTS

Greenwood Cotton Insulation Products, Inc.
P.O. Box 1017, Greenwood, SC 29648; (800) 546-1332 or (404) 998-8888
Kraft-faced cotton insulation.

Insulcot Cotton Insulation Products
70 Munsell Court, Roswell, GA 30076; (404) 998-6888
Made from 75 percent recycled cotton fibers and 25 percent polyester, which binds the fibers together.

Ideal Environments
401 W. Adams, Fairfield, IA 52556; (515) 472-6547
Distributor of Insulcot.

Environmentally Safe Products, Inc.
313 West Golden Lane, New Oxford, PA 17350; (800) BUY-LowE or (717) 624-3581
Manufacturers of Low-E Insulation, made from polyethylene foam, about 20 percent of which is recycled material, with aluminum backing. Can be used for various applications for which many insulations are inappropriate, including water-heater blankets, duct and pipe wrap, and air-conditioner covers.

BLOWN-IN INSULATION
Greenwood Cotton Insulation Products, Inc.
P.O. Box 1017, Greenwood, SC 29648; (800) 546-1332 or (404) 998-8888
Cotton insulation for blown-in applications.

FOAMED-IN INSULATIONS
Air Krete, Inc.
P.O. Box 380, Weedsport, NY 13166; (315) 834-6609
Air Krete is a cementitious compound typically made from magnesium oxide, a mineral derived from seawater. The best alternative to sprayed-in-place synthetic foams for filling concrete blocks, brick cavities, ceilings and floors, and around pipes. Best used in existing houses; can be used in new construction but will increase costs. R-value higher than fiberglass or cellulose. Does not shrink and maintains high R-value. Unlike synthetic foams, does not burn. Competitive in price with other foamed-in insulations. However, Air Krete is not widely available and must be installed by a trained technician.

Nordic Builders
162 North Sierra Court, Gilbert, AZ 85234; (602) 892-0603
Air Krete insulation.

Palmer Industries
10611 Old Annapolis Road, Frederick, MD 21701; (301) 898-7848
Air Krete insulation.

RIGID INSULATIONS
Synthetic foam boards blown with less-ozone-depleting HCFCs are appropriate for all situations in which other insulations are unsuitable or greater R-value per inch is required. They also include some recycled material.

Amoco Foam Products Company
400 Northridge Road, Suite 1000, Atlanta, GA 30350; (800) 241-4402 or (404) 587-0535
Amofoam-RCY is blown with HCFC-142b, which has an ozone depletion potential 94 percent less than that of the CFC-12 it replaced. Contains a minimum of 50 percent recycled polystyrene, both consumer discards and industrial scrap.

CAULKS AND SEALANTS

Caulks and sealants have many uses around the home. One of the most important is conserving energy. The average home has more than two hundred square inches of cracks and gaps that are believed to be the cause of 20 to 50 percent of total household heat loss. This means that the unwanted air that sneaks into your house can account for as much as half of your heating and cooling bills! Fortunately, the tools to fight air leaks are inexpensive: caulks, sealants, gaskets, and weatherstripping.

First, some definitions. Technically, a caulk is an asphaltic (coal-based) or linseed oil–based compound, whereas a sealant is a synthetic plastic resin. However, in the industry, the terms are often used interchangeably. A gasket is a piece of string, rubber, foam, paper, or other material placed around an object or joint to make it leakproof. Weatherstripping is a material or combination of materials that form a barrier to prevent air from leaking in through

As a result of the energy crises of the 1970s a building technique known as airtight drywall approach, or ADA, became popular. Almost all joints — for example, sill to foundation — are gasketed or sealed.

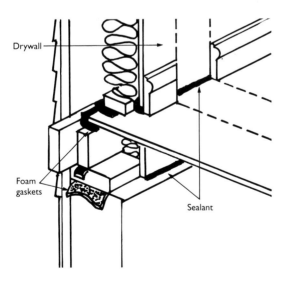

gaps, especially around doors and windows.

This section will deal primarily with caulks and sealants. A dizzying array of these products is now on the market, with over two hundred manufacturers in the United States alone. Yet sealants must be chosen carefully, because of all the products used to weatherize a home, they have the greatest potential impact on indoor air quality, and hence your health.

Conventional Products

Asphaltic and linseed oil caulks were the only kinds available until the 1940s, when latex and other plastic-resin sealants were first developed. These new synthetic products, which perform better than traditional caulks for most uses, have taken over the market. Today, there are so many that the hardest part of a caulking job is choosing which to buy. Among the most commonly used categories of sealants are acrylic latex, butyl, oleoresinous, silicone, solvent acrylic, and polyurethane.

Health and Environmental Concerns

Many of the chemicals used in the manufacture of sealants are potentially toxic. Because there are hundreds of products on the market, and the precise composition of each is proprietary information, it's impossible to list all of the constituent chemicals for every brand and their potential health effects. Suffice it to say that toluene, xylene, and ethylene dichloride are among the toxic substances found in sealants, some of them suspected cancer causers. This is of particular concern when caulks and sealants are used indoors. Some of the toxic constituents of sealants volatilize, or become a breathable gas, at room temperature. These volatile organic compounds (VOCs) can add substantially to the pollutant load in your indoor air for weeks or months (see "A Guide to Home Pollutants," pages 245–249).

Some sealants have a more adverse effect on indoor air quality than others. A study con-

LEAK DETECTORS

In the 1970s, caulking and weatherstripping became household terms. During those years, when the price of energy went through the roof, we plugged up air leaks around doors and windows and in all the other obvious places. But those leaks are just the tip of the iceberg. One study showed that a simple test with a device called a blower door can detect leaks that, when eliminated, result in energy savings seven times greater than the savings achieved when homeowners detect and plug leaks themselves.

Here's how the test works: a tight-fitting door is temporarily installed in the doorway of a major entrance to your home. This blower door has a variable-speed fan and pressure gauge built in. With windows, doors, and fireplace dampers closed, the blower-door fan pulls air outside the house, depressurizing it. By measuring how hard the fan must work to maintain the difference in pressure between the inside and outside air, the blower door technician can tell you how tight the house is. If your house fails to meet recommended air-tightness standards, he or she will proceed to walk through your rooms, searching with a harmless "smoke wand" for air infiltrating not only through window and door frames but also electric wall switches, cracks in the wall, warped baseboards, and a variety of other hard-to-find places. Leaks are detected when the smoke veers toward the blower door. You'll notice the difference in your energy bills when you seal up these leaks.

Many utilities offer these tests as a service to their customers. If yours doesn't, you can hire an energy auditor to conduct the blower door test.

---

ducted for the Canadian government examined the safety of various sealants for indoor use, based on tests of their ability to emit potentially toxic air pollutants. The study rated the following sealants as either safe for typical indoor use, safe in very limited quantities for indoor use, or unsafe for indoor use:

Safe for typical indoor use:
Oleoresinous
Acrylic latex
Polyurethane
Silicone
Polysulfide

Safe in very limited quantities:
Solvent-based acrylic
Butyl rubber
Neoprene

Unsafe:
Styrene butadiene
Nitrile

Environmental Products

Because caulks and sealants play such a unique and important role in conserving energy in the home and consequently reduce the environmental pollution resulting from energy production, and because few products made of natural or low-VOC ingredients are available, this is an instance in which it is often necessary to choose the lesser of two evils. If you're building a new home, you can avoid sealants to a large extent by designing a structure that minimizes gaps and leaks. For example, after the energy crises of the 1970s a building technique known as the airtight drywall approach became popular. It relies heavily on the use of foam gaskets. Almost all joints — sill to foundation, band joist to sill, subfloor to band joist, wall plate to subfloor, drywall to frame — are gasketed. Adobe, rammed earth, and other houses constructed of one seamless material also minimize air leaks. Short of building a new home, however, there are some things you can do to minimize your exposure to pollutants as you plug up air leaks:

- Use only as much sealant as you need to get the job done.
- Even in an existing house, you can use foam tape gaskets instead of sealants for some applications. The foam tapes are best suited for larger openings, such as those around windows and doors, and so-called expansion joints where wood meets concrete, masonry, or metal.
- For smaller, hairline joints or where a waterproof seal is required, you'll still need a caulk or sealant. Examples are where a countertop, bathtub, or shower meets the wall. Read the label carefully, and use only the safest sealants listed above. Water-based latex products are generally less hazardous than those that are solvent-based.
- Natural caulks made of wood resins, plant gums, and the like are as effective as the synthetic sealants for caulking interior cracks and joints and are definitely worth considering.
- For other uses, look for one of the new low-VOC synthetic sealants.

Suppliers

NATURAL SEALANTS
Linseed Putty, Eco Design
1365 Rufina Circle, Santa Fe, NM 87501; (800) 621-2591 or (505) 438-3448

Natural Sealant #386, Auro, Sinan Co.
P.O. Box 857, Davis, CA 95617-0857; (916) 753-3104

LOW-VOC SEALANTS
Dyno Flex Caulking Compound, AFM Enterprises, Inc.
1960 Chicago, Suite E7, Riverside, CA 92507; (909) 781-6860

GL700 Silicone Acrylic Caulk and GL600 Acrylic Latex Caulk, Green Line. Available through **Environmental Construction Outfitters**
44 Crosby Street, New York, NY 10012; (800) 238-5008 or (212) 334-9659

Phenoseal Vinyl Adhesive Caulk, Gloucester Company
P.O. Box 428, Franklin, MA 02038. Has been used by chemically sensitive people for years. For a dealer near you, call (800) 343-4963.

W. F. Taylor
13660 Excelsior Drive, Santa Fe Springs, CA 90670; (213) 802-1896
Low-VOC sealants.

FOAM TAPE GASKETS
Sure Seal and Ultra Seal, Denarco Sales Company
12710 Idlewilde Street, White Pigeon, MI 49099; (616) 641-2206

Will-Seal, Illbruck/USA
3800 Washington Avenue North, Minneapolis, MN 55412; (612) 521-3555

Preservatives have been around almost as long as wood has been used as a building material. The American colonists dipped wood in a solution of arsenic salts to protect it against attack by insects. Creosote, derived from coal tar, was the first wood preservative effective in moist areas and where wood comes in contact with the soil. Patented in 1831, it was the most widely used preservative for many years. Pentachlorophenol, the first synthetic preservative, was developed in the 1930s.

Today, about 3.8 billion board feet of treated lumber is used in North America every year. According to the American Wood Preservers Association, treated wood accounts for more than 21 percent of all the dollars spent on timber products. Around the home treated wood is used for fence posts, arbors, raised beds, outdoor furniture, and especially decking.

Conventional Products

In 1986, the U.S. Environmental Protection Agency restricted the sale of the three major wood preservatives, creosote, pentachlorophenol, and inorganic arsenic compounds, to trained and certified workers. These wood preservatives can no longer be sold to homeowners, although lumber and other wood products that have already been treated with them are still available.

The preservative used on most lumber sold for home use is an inorganic arsenic compound, chromated copper arsenate, or CCA. The copper keeps fungi from attacking the wood; the arsenic holds off termites and other wood-chewing insects; and the chrome binds the preservative to the lumber. CCA-treated wood is the stuff at home centers and lumberyards that has a greenish cast. It's called pressure-treated wood because in order to be effective the wood must not only be impregnated with preservative but also subject to high pressure in

cylindrical chambers to force the preservative into the wood cells. Products available for use by homeowners who wish to treat wood themselves typically contain copper naphthenate, which is somewhat less toxic than CCA.

Health and Environmental Concerns

Toxicity

Chromated copper arsenate is toxic, there's no doubt about that, or its sale wouldn't be restricted by the EPA. The agency forbids its use in products where it may come into direct or indirect contact with drinking water or food — cutting boards, countertops, or picnic tables, for example. It allows use of the preservative indoors as long as all sawdust and construction debris is cleaned up and disposed of when construction is completed. Because for homeowners the primary health risks are ingesting or inhaling sawdust, the EPA also instructs users to wear goggles and dust masks when sawing pressure-treated wood, to wash thoroughly before eating or drinking, and to launder work clothes separately before using them again. In addition, the agency warns against using treated wood that shows evidence of CCA surface precipitates (usually a white crystalline residue), which can be absorbed through the skin on contact. Because the raw chemicals are ten to one hundred times more concentrated before they are diluted and applied to wood, the health and environmental risks are much greater during manufacture and transport and at the plants where the wood is treated.

Disposal

Pressure-treated wood can last fifty years or more. While this is its strength, it is also its curse — there is no environmentally sound way to dispose of treated wood once it has outlived its use. This is destined to become a major problem as people begin to remodel and relandscape homes where so many billion board feet have been used in recent years. Treated wood should never be burned outdoors or in wood stoves because toxic metals will be released directly into the air. In landfills, it takes up valuable space and won't break down for many decades. If it is burned in municipal incinerators with state-of-the-art pollution control equipment, the heavy metals will not be released into the air but will end up in the incinerator ash, from which it can leach into ground or surface waters when deposited in landfills.

Environmental Products

Even if you use treated wood according to EPA directions and avoid lumber with a white surface residue, you're still supporting a huge industry that produces and transports large amounts of hazardous substances, and you're contributing to a waste problem that has the potential to be a major headache in years to come. There are a number of environmental alternatives:

- Whenever possible, use alternatives to wood. For example, a stone patio is just as handsome as a wooden deck. Locally quarried stone is recommended because it doesn't have to be transported long distances, saving energy.

- Avoid the need for preservatives by using naturally rot-resistant woods from certified foresters (see the list of rot-resistant woods on page ooo). But be sure to use only the heartwood of even these species, because the sapwood has little or no decay resistance.

- Use salvaged redwood, cypress, and other species that have been the traditional woods of choice when rot resistance is necessary. This relieves some of the pressure to harvest remaining old-growth stands of these trees. Salvaged old-growth wood is also superior to second-growth timber, which has wider growth rings and is therefore less dense and decay resistant.

- Consider recycled-plastic lumber, which not only provides a use for discarded plastics but also is impervious to decay (see "Recycled-Plastic Lumber," pages 137–140).

- Use woods that are not naturally rot resistant and protect them with a natural or low-pollutant paint or finish (see "Enamel Paints," pages 185–187, and "Sealers and Finishes," pages 189–197).

- Use borate preservatives, derived from boron, a naturally occurring mineral that's used in hundreds of products from fertilizers to eyedrops. Borates are relatively nontoxic and are effective wood preservatives, but they can leach out of wood that is exposed to water. Where the wood remains dry, borate preservatives are an excellent choice to prevent damage from termites and other insects. Exposed woods must be treated with a borate preservative and then coated with a water-resistant finish. Borate preservatives are available as pellets that are inserted into the core of the wood and liquids that are brushed or sprayed on.

Suppliers

Nisus Corporation

Cherokee Place, 101 Concord Street North, Knoxville, TN 37919; (800) 264-0870
Impel Rods and Bora-Care liquid borate preservatives.

Perma-Chink Systems, Inc.

If you live in the West, write the Western Division at 17635 N.E. 67th Court, Redmond, WA 98052; or call (800) 548-1231. In the East, write the Eastern Division, 1605 Prosser Road, Knoxville, TN 37914; or call (800) 548-3554
Bora-Care and TimBor borate preservatives.

Although more steel is used annually in the United States for construction than for any other purpose — 11 million tons, compared to 10 million for automotive production — until recently little was used to build homes. However, high timber prices, particularly on the West Coast, have created a strong market for steel-framed production homes. According to Western Metal Lath, a supplier of metal studs based in Riverside, California, demand for steel studs for homes shot up 500 percent in 1992. Many of the companies that sell production homes offer wood and steel framing for the same price; most buyers select steel over wood.

The typical home requires five to six tons of steel. An estimated 8 million tons would be required to meet the needs of the U.S. homebuilding industry. Steel manufacturers see this as an opportunity, not a problem: with automobiles being built increasingly of plastic and aluminum parts to reduce weight, the industry needs to find new markets. Right now, most steel-stud manufacturing plants are operating well below capacity.

Health and Environmental Concerns

Resource Depletion

Steel is not a renewable resource. It is made of iron ore, nickel, chromium, manganese, and other minerals that are in limited supply. What's more, large quantities of water are used in the manufacture of steel.

Energy

Compared to wood, steel production requires enormous amounts of energy — approximately 22 million Btus per ton of end product. Recycling steel in electric furnaces consumes about 6.3 million Btus per ton, or about 39 percent of that required to produce it from raw materials.

Mining Impacts

Mining the minerals required for steel production causes soil erosion and air and water pollution. Land is disturbed and habitat lost.

WOOD STUDS VERSUS STEEL STUDS
An Environmental Comparison

WOOD	STEEL
Amount required to frame the average home: about 9,100 board feet, or 40 to 70 trees	Amount required to frame the average home: 5 or 6 tons
Renewable resource, but much comes from threatened old-growth forests	Iron ore, chromium, zinc, nickel and limestone are nonrenewable resources
Salvageable	The most recyclable building material
Not energy intensive to manufacture	Very energy intensive to manufacture
Little pollution produced during manufacturing	Air and water pollution produced during manufacturing and recycling
Biodegradable	Non-biodegradable
Does not conduct heat	Conducts heat
Emits natural terpenes	Emits no pollutants

Air Pollution

Steel manufacturing results in dust, particulates, and combustion emissions, including carbon dioxide, a major greenhouse gas, and sulfur dioxide, a cause of acid rain.

Environmental Products

As a framing material, recycled steel is superior to wood in many respects. It is resistant to termites, rot, and UV rays, and hence requires no sealers or preservatives. It's stronger than wood, withstanding 50,000 pounds per square inch versus 36,000 psi for a conventional two-by-four, and performs four times better than wood during earthquakes. It doesn't warp, bend, twist, or bow. No waste is produced at the construction site; typically, about 25 percent of wood must be discarded due to defects such as warping or knots. Steel doesn't absorb moisture, doesn't check or crack. And it exhibits little thermal expansion and contraction.

Steel is also a lot lighter than wood and is therefore not only easier to work with but also less energy intensive to transport. Fewer studs, joists, and rafters are required. Steel suffers less damage than wood when foundations settle. Because it is inert and emits no irritants or pollutants, it is often used in homes for the chemically sensitive. Using steel results in less noise at the construction site because power saws and hammers are unnecessary (screws, not nails, are used with steel studs). Steel is more durable than wood, and virtually maintenance free. And perhaps most important of all, steel is noncombustible.

Although steel is 100 percent recyclable, steel studs currently consist of varying percentages of recycled material. Using steel for home building makes sense environmentally only when it is composed of a substantial percentage of recycled material. If you decide to frame your home with steel, look for studs with the highest recycled content.

Suppliers

American StudCo
2525 North 27th Avenue, Phoenix, AZ 85009; (800) 877-8823

Angeles Metal Systems
4817 East Sheila Street, Los Angeles, CA 90040; (213) 268-1777

Approved Equal
1538 Gladding Court, Milpitas, CA 95035; (408) 942-8191

California Building Systems
4817 East Sheila Street, Los Angeles, CA 90040; (213) 260-5380

CEMCO
263 Covina Lane, City of Industry, CA 91744; (818) 369-3564

John W. Hancock, Jr., Inc.
Diuguids Lane, Salem, VA 24153; (703) 389-0211

Knorr Steel Framing Systems
5073 Salem Dallas Highway, P.O. Box 5267, Salem, OR 97304; (503) 371-8033

Steeler
1003 Martin Luther King Way South, Seattle, WA 98178; (206) 725-2500

U.S. Posco
900 Loverdge Road, Pittsburg, CA 94565; (510) 439-6000

Vulcraft
A division of NUCOR Corporation, has five plants across the country:

Alabama: P.O. Box 169, Fort Payne, AL
35967; (205) 845-2460
Indiana: P.O. Box 1000, St. Joe, IN 46785;
(219) 337-5411
South Carolina: P.O. Box F-2, Florence, SC
29502; (803) 662-0381
Nebraska: P.O. Box 59, Norfolk, NE 68701;
(402) 371-0020
Texas: P.O. Box 186, Grapeland, TX 75844;
(713) 687-4665

Western Metal Lath
P.O. Box 39998, Riverside, CA 92159; (909)
360-3500

RECYCLED-PLASTIC LUMBER

The advent of plastic lumber marks the advent of the depletion of forests. Another reason for the development of recycled-plastic building products is that they provide a use for the estimated 9 million pounds of plastic waste that end up in landfills every year; 99 percent of the 10.5 million pounds of plastic generated annually in the United States is not recycled.

Unlike many plastic manufacturing processes that require extremely pure polymer resin, plastic lumber can be made with commingled, or mixed, plastics. Consequently, recycled-plastic lumber provides a use for waste material that the plastics industry would otherwise be unable to use. High-density polyethylene (HDPE), of which detergent containers and many other plastic bottles are made, is the resin most commonly used by manufacturers of plastic lumber. Some recycled-plastic lumbers are a mixture of plastic and other material such as glass fiber or wood.

Health and Environmental Concerns

In a perfect world, plastics would not be used frivolously in so many disposable products but rather reserved for those uses for which there are few alternatives. Plastics are made of nonrenewable fossil fuels, and manufacturing them results in extremely toxic, and highly regulated, by-products and emissions. They are also extremely high in embodied energy, meaning that a great deal of energy is needed to produce them.

Recycling plastics into lumber raises some comparatively minor concerns. If waste water from washing the recycled materials is generated, it must be disposed of properly or it can contaminate ground and surface water. The wide variety of additives used in making plastic lumber — UV stabilizers and anti-oxidants to retard degradation over time, blowing agents to create lighter-weight material, compatibilizers to allow for some degree of bonding between plastic polymers that don't normally bond — is another possible concern, although experts say that they are used in relatively minute quantities. In addition, recycled-plastic lumber that has been chemically cross-linked is not recyclable into lumber or any other material.

Environmental Products

Diverting millions of pounds of plastic waste from overburdened landfills is the biggest benefit of using recycled-plastic lumber. As a replacement for treated wood, it has some important advantages. Pressure-treated wood is saturated with toxins (see "Wood Preservatives," pages 133–134). Even pressure-treated wood is still vulnerable to attack from marine borers. Disposal of used treated wood is a growing problem, and the list of states requiring that it go to hazardous waste landfills is growing. By comparison, recycled-plastic lumber is insect and rot resistant and virtually maintenance free. It has many potential uses around the home: fencing and gates, steps and risers, mailbox posts, outdoor furniture, sandboxes and playsets, storage, compost, and trash bins, landscaping timbers, retaining walls, arbors and trellises and planters. At waterfront homes, recycled-plastic lumber can also serve as pilings, boardwalks, docks, and ramps. On farms or ranches it can be used for crop and fruit-tree supports, corrals, and animal pens.

Be forewarned, though, that the plastic-lumber industry is in its infancy, and there are as yet no testing or industry standards by which products can be compared. Indeed, the term "recycled-plastic lumber" encompasses a broad range of materials with various qualities and characteristics. For example, lumber made from both wood and plastic wastes looks a lot more like natural wood than the solid plastic products. Recycled-plastic lumber is also more ex-pensive than wood. Because many of these materials have significant weaknesses, here are some things to keep in mind:

- Choose a product suited to your application. Plastic lumber made from unreinforced commingled plastic has low strength-to-weight ratio and poor span ratings compared to wood or lumber made from pure HPDE. Highly foamed or hollow extruded products are stronger than solid plastic materials. Where structural strength isn't necessary, use inexpensive commingled-plastic products manufactured locally.

- If the lumber will be exposed to direct sunlight, test a sample in the sun to make sure that it doesn't become uncomfortably hot to the touch.

- Unless you're using products with large percentages of wood or glass fiber, make allowance for expansion and contraction; experts recommend about $1/4$ inch per 8-foot length.

- Fasten plastic lumber together with through-bolts or screws, not nails.

- Use carbide-tipped tools, not high-speed steel, to work plastic lumber.

Suppliers

Aeolian Enterprises, Inc.
1 Lloyd Avenue, Latrobe, PA 15650; (412) 539-9460

AERT
P.O. Box 1237, Springdale, AR 72765; (501) 750-1299

Aldan Lane Company
P.O. Box 990, Kalona, IA 52247; (319) 656-3620

Bedford Industries, Inc.
1659 Rowe Avenue, P.O. Box 39, Worthington, MN 56187; (507) 376-4136

BTW Industries, Inc.
2000 SW 31st Avenue, Pembroke Park, FL 33009; (305) 962-2100

Carrysafe
920 Davis Road, Suite 101, Elgin, IL 60123; (800) 231-9721

Coon Manufacturing
P.O. Box 190, Spickard, MO 64679; (816) 485-6299

Custom-Pac Extrusions, Inc.
16865 Park Circle Drive, Chagrin Falls, OH 44023; (216) 543-8284

Cycle-Masters, Inc.
P.O. Box 1864, Elkhart, IN 46515; (219) 293-5636

Duratech Industries, Inc.
1842 W. Easton Highway, Lake Odessa, MI 48849; (616) 374-0240

Eaglebrook Products, Inc.
2600 W. Roosevelt Road, Chicago, IL 60608; (312) 638-0006

Earth Care Products, Inc.
P.O. Box 5937, Statesville, NC 28687; (704) 878-2582

Earth Safe
P.O. Box 2861, Hyannis, MA 02601; (508) 420-5681

Eco-Tech LP
4004 Dayton, McHenry, IL 60050; (815) 363-8570

Environmental Recovery Systems
1400 Brayton Point Road, Somerset, MA 02725; (508) 677-0252

Environmental Recycling, Inc.
8000 Hall Street, St. Louis, MO 63147; (314) 382-7766

Envirowood Corp.
501 W. Algonquin Road, Mt. Prospect, IL 60056; (800) 323-0830

Futurewood Products
934 Easton Street, Ronkonkoma, NY 11779; (516) 588-4545

Green Tree Plastics Technology
22073 Loch Lomand Drive, Canyon Lake, CA 92587; (909) 244-4515

Hammer's Plastic Recycling Corp.
RR 3, Box 182, Iowa Falls, IA 50126; (515) 648-5073

Innovative Plastic Products
P.O. Box 898, Greensboro, GA 30642; (706) 453-7552

International Plastics Corp.
301 E. Vine Street, Suite B, Lexington, KY 40507; (606) 233-4332

Iowa Plastics, Inc.
3464 Goldfinch Avenue, Hull, IA 51239; (712) 722-0692

Jeanell Sales Corp.
P.O. Box 537, Sharon Industrial Park, Sharon, TN 38255-0537; (901) 456-2681

Mobile Oil Corporation
800 Connecticut Avenue, Norwalk, CT 06856; (203) 831-4200

NEW Plastics Corp.
P.O. Box 220, Luxemburg, WI 54217; (414) 845-2326

Partek Corp.
P.O. Box 1387, Vancouver, WA 98666; (206) 695-1777

Plastic Pilings, Inc.
8560 Vineyard Avenue, Suite 205, Rancho Cucamonga, CA 91730; (909) 989-7685

Plasticera, Inc.
3417 E. Columbus, Tampa, FL 33605; (813) 248-2559

Plastic Lumber Co., Inc.
540 South Main Street, Akron, OH 44311-1010; (216) 762-8989

Recycled Plastics Industries, Inc.
1820 Industrial Drive, Green Bay, WI 54302; (414) 468-4545

Recycled Polymer Associates
152 W. 26th Street, New York, NY 10001; (212) 463-8622

St. Jude Polymer Corp.
1 Industrial Park, Frackville, PA 17931; (717) 874-1220

Sanders Enterprises, Inc.
3019 Nash Road, Scott City, MO 63780; (314) 334-9600

Superwood of Alabama, Inc.
P.O. Box 2399, Selma, AL 36702; (205) 874-3781

Trimax of Long Island
2076 5th Avenue, Ronkonkoma, NY 11779; (516) 471-7777

United Resource Recovery
Route 2, Box 265, Jonesboro, AR 72401; (501) 932-3500

Westmont Building Products Co.
200 E. Quincy, Westmont, IL 60559; (708) 968-3420

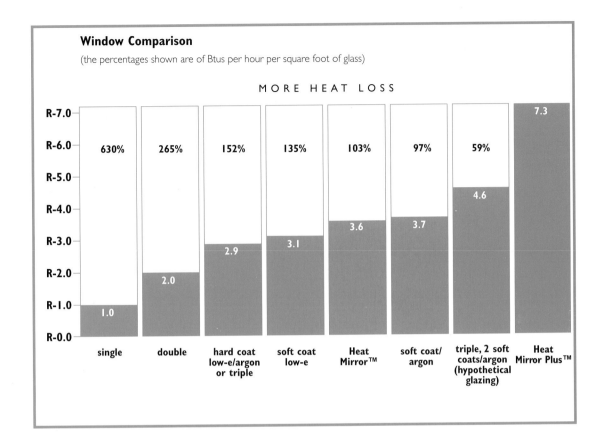

Window Comparison

(the percentages shown are of Btus per hour per square foot of glass)

MORE HEAT LOSS

	single	double	hard coat low-e/argon or triple	soft coat low-e	Heat Mirror™	soft coat/ argon	triple, 2 soft coats/argon (hypothetical glazing)	Heat Mirror Plus™
%	630%	265%	152%	135%	103%	97%	59%	
value	1.0	2.0	2.9	3.1	3.6	3.7	4.6	7.3

WINDOWS

A profusion of glass affords modern-day humans, in the words of Frank Lloyd Wright, "something of the freedom of our arboreal ancestors living in their trees." Windows blur the distinction between nature and architecture, enabling us to live in close touch with the natural world. Windows allow daylight into the home, reducing the amount of electricity needed for artificial lighting. Windows also capture breezes, helping to ventilate a home naturally.

It's no wonder, then, that during the 1980s windows were all the rage. Bay and bow windows were de rigueur in remodelings. The Palladian window, a tall window with an arched top and long rectangular windows on either side, was popular in remodelings as well as in new construction. The transom window re-

turned. Bedroom skylights revealed constellations in the nighttime sky.

Health and Environmental Concerns

Energy Efficiency

Alas, the twentieth-century concept of bringing the outdoors inside with transparent walls collided with another critical consideration — energy. The intellectual achievement represented by glass walls in houses and skyscrapers could be sustained only temporarily by the availability of cheap fossil fuels. Precious energy resources were wasted as heat passed easily through the large panes of glass.

Twenty to 30 percent of an average home's heat is lost through its windows; it has been estimated that the amount of energy flowing out of America's windows is equivalent to the amount of oil flowing through the Alaska pipeline — 1.8 million barrels a day.

Glazing, or the panes of glass and the way they're sandwiched together in the window sash, is more important than any other factor in determining a window's thermal efficiency. Heated air is sucked right out of a home with windows consisting of the old-fashioned single pane of glass; single-pane glass rates a miserable 1 on the R-value scale. (R-value, whether in insulation or in windows, is the measure of a material's ability to keep heat from flowing into or out of a room.) Today, in all but the mildest climates, double glazing is the norm. Double glazing usually consists of two panes of glass with a hermetically sealed airspace in between. When it traps this layer of air between the panes, double glazing is called insulating glass and achieves an R-2 rating.

Resource Depletion

The use of some wood species, such as mahoganies, in the construction of window sash has contributed to unsustainable forestry practices in tropical and domestic forests (see "Wood Products," pages 97–120).

The virgin aluminum used in window construction is extremely energy intensive to produce. In addition, the bauxite from which aluminum is made is a finite resource, and mining it destroys tropical forest habitat.

The vinyl used increasingly in window construction is made from nonrenewable fossil fuels.

Toxicity

Vinyl manufacturing results in extremely toxic and highly regulated wastes.

Indoor Air Quality

The sealants used in glazing are all petrochemical formulations that can pollute the air inside your home (see "Caulks and Sealants," pages 130–132.) Although there is no extreme health hazard associated with the offgassing of these sealants — or of that from the vinyl or wood preservatives and coatings used in window assemblies — each of these sources adds to the load of irritants and toxins released to the indoor air by the hundreds of building, decorating, and maintenance products used in the typical home.

Environmental Products

A new generation of energy-saving windows is slashing the amount of fuel wasted by single or double glazing. We're fast approaching the day when windows are as good as walls — or better, because they enable us to open our cocoon-like shelters to the sunlight.

In the early 1980s the first low-E, or low-emissivity, coatings were developed. The beauty of a low-E coating, an atoms-thin layer of metal typically applied to one of the panes in a double-glazed window, is that it reflects heat but lets through almost all incoming light, with the exception of most ultraviolet rays that damage fabrics. In cold areas, heat trying to escape is reflected back into the house, increasing the window's energy efficiency to approximately R-3. Low-E windows make sense not only in cold regions but also in warm ones. In hot areas, a window's R-value is often less of a concern than its "shading coefficient," the measure of its ability to transmit solar heat. Low-E windows have the lowest possible shading coefficients because the coating deflects the sun's warming rays before they can enter the house.

Even better is a low-E window with argon gas instead of air in the space between the panes. Argon conducts heat less readily than air. It is an environmentally safe, cheap, and widely available gas that comprises about one percent of the earth's atmosphere. Filling a low-E window with argon can boost its energy rating to R-4 or more. Unfortunately, the argon gas will escape over time, twenty years or so, reducing the window's original energy efficiency.

Windows with Heat Mirror instead of low-E-coated glass panes are more energy efficient still. Developed by Southwall Technologies of Palo Alto, California, in 1981, Heat Mirror consists of a thin plastic film with a low-E coating. Stretched inside the airspace between two panes of glass, it effectively creates two insulating airspaces. Double glazing with argon gas and Heat Mirror boosts energy efficiency even more. There are a variety of different types of Heat Mirror, each designed for a specific climate or use. Window manufacturers combine them in different ways. For example, Alpen, Inc., in Boulder, Colorado, produces a window composed of two layers of Heat Mirror suspended between two panes of glass with krypton gas between the panes. (Krypton, another environmentally safe gas, conducts even less heat than argon). The window looks like ordinary double glazing but with an R-10 rating is five times more energy efficient.

Cloud Gel, developed by Suntek in Albuquerque, New Mexico, is an even more recent innovation in glazing. A specially formulated polymer mixed with water to create a gel, it turns milky white when the window reaches a certain temperature, thereby blocking out additional undesired heat. Although heat is reflected, light is not. Consequently, Cloud Gel saves energy that would otherwise be used for cooling as well as artificial lighting. With a rating of R-13, double glazing with krypton gas, Heat Mirror, and Cloud Gel is as energy efficient as a wall filled with 3½ inches of fiberglass insulation!

The window industry is hard-pressed to keep pace with the breakthroughs in glazing technology. It won't be too long, for example, before so-called electrochromic or "smart" glazings are widely available. When electricity is applied to electrochromic glass, a coating on the pane darkens so that it transmits less heat and light.

Photovoltaic or electric-generating glazings, which actually *produce* usable electricity while preventing heat loss or excess solar heat gain, are also being developed.

Environmentally sound windows aren't just energy efficient. Ideally, they're also made of wood from certified sustainable sources (see pages 106–107). Salvaged windows with the old panes replaced with energy-efficient glazings are another option. Some manufacturers are also conserving resources by using recycled plastics or aluminum in the window assembly. In addition, they're designing windows that are easily disassembled for recycling in the future. To keep the air inside your home healthy, the best windows are also finished with natural or low-pollutant paints or finishes.

There are a few other things to consider when shopping for windows. Both the frame and sash play a substantial role in a window's thermal efficiency. The sash consists of the vertical and horizontal pieces of wood, metal, or plastic into which the glass is set. The sash fits into the frame, which comprises the jambs and sill that surround the window opening. Manufacturers tend to tout the R-value of the glazing alone when they talk about the energy efficiency of their windows. Yet the R-value of the glass may be quite different from the R-value of the window as a whole. Be sure to ask about the overall R-value of the window assembly.

And don't just look at the ticket price when you're shopping for windows; be sure to factor in the future energy savings that the various kinds of insulating windows will accrue. Although the initial cost of a window may seem high, savings on heating and cooling bills can quickly recover the higher initial cost — and actually save you money from then on.

Suppliers

Almost every major window manufacturer offers models with low-E coatings and argon gas. A few offer windows incorporating Heat Mirror.

Architectural Openings, Inc.
16 Garfield Avenue, Somerville, MA 02145; (617) 776-9223
Windows constructed with certified woods and finished with natural or low-toxicity coatings. The windows are also easily disassembled for recycling. All are made with Heat Mirror and argon gas.

Hope's
84 Hopkins Avenue, P.O. Box 580, Jamestown, NY 14702-0580; (716) 665-5124
All steel windows. Some parts made with recycled steel. Low-E only.

Hurd Millwork Company, Inc.
575 South Whelen Avenue, Medford, WI 54451; (715) 748-2011
Windows with Heat Mirror.

Southwall Technologies
1029 Corporation Way, Palo Alto, CA 94303; (800) 365-8794 or (415) 962-9111
Manufacturers of Heat Mirror Insulating Glass. Write or call for a list of window companies that use Heat Mirror.

Suntek, Inc.
6817 Academy Parkway East, Albuquerque, NM 87109; (505) 345-4115
Manufacturer of Cloud Gel.

LIFE-SUPPORT
SYSTEMS

HEATING AND COOLING

Jolted by the energy crises of the 1970s, we caulked and weatherstripped, stuffed our walls with insulation, and began paying attention to the amount of energy it takes to run our appliances. As a result, the upward-spiraling consumption of energy in American homes and buildings fell about 30 percent.

But there's still plenty of room for improvement; in fact, we're headed in the wrong direction. In 1986, when the price of oil plummeted, the movement toward energy efficiency stalled, and energy use now is once again on the upswing. Ironically, it's easier than ever to have a home that conserves energy. New technologies, from superwindows to computer-run heating and cooling systems, combined with time-tested strategies such as shade trees and insulation, could further slash the heating and cooling bills run up by American homes.

Most homeowners balk at making these improvements because they cost money. But rather than being deterred by the costs of energy improvements, you should view them as attractive investment opportunities. Energy experts Rick Bevington and Arthur Rosenfeld make this point by comparing the cost of two new homes in Chicago: one has conventional

walls, with standard 3½-inch insulation. The other has thicker walls and six inches of insulation. Although the extra insulation cost $300, the heating bill for the more efficient house drops approximately $120 a year. That means the payback period for the insulation is about two and a half years — which translates into an annual return on investment of 40 percent!

Conventional Products

Most American homes are heated with either oil or natural gas. In remote locations, wood or propane are often used. Some areas depend primarily on electric resistance heating such as electric baseboard systems, though these tend to be expensive. In mild climates, air-source heat pumps, which extract heat from one place and release it in a warmer place, are employed for heating in winter and cooling in summer. The vast majority of Americans use air conditioners to cool their homes. Although the efficiency of many models has improved, air conditioners still gobble up a lot of energy, and to add insult to injury, they rely on ozone-depleting coolants.

Together, these heating and cooling systems consume a mind-boggling amount of energy, almost 60 percent of the energy used in American homes, and so they're a good place to aim for savings.

Health and Environmental Concerns

Most of us take energy for granted. Sure, we think about it for about as long as it takes to pay the monthly utility bills. But precious few of us associate our furnaces or air conditioners with the strip mines, oil rigs, and nuclear generating stations that power them. A thorough look at the environmental impacts of home energy use could easily fill volumes. For starters, consider the fact that each form of energy we consume involves a multitude of deleterious environmental effects from the moment it is extracted.

Extraction

Coal is often strip-mined. Oil drilling disturbs sensitive wetland and marine habitats. Radioactive tailings from uranium mines threaten the health of miners and contaminate the surrounding land.

Manufacturing

Home heating oil must be refined from petroleum stocks. Liquified petroleum gas (propane), kerosene, and other fuels are also created out of crude oil. Uranium ore must be processed for use in nuclear power plants. All of this results in considerable air and water pollution and hazardous wastes.

Transportation

We've all witnessed the tremendous toll on the environment wrought by spills in Alaska's Prince William Sound and around the world as oil is shipped from place to place. The transport of nuclear feedstocks and wastes is a potentially lethal business. What's more, a great deal of fuel is needed to transport these fuels around the globe.

Combustion

The combustion of fossil fuels in our homes and at power plants results in emissions of sulfur dioxide and nitrogen oxides, which cause acid rain. And the energy that powers our appliances and heats, cools, and lights the homes and buildings in the United States produces 500 million tons a year of carbon dioxide, a gas that causes the greenhouse effect, which many scientists believe will inevitably lead to global warming.

Disposal

Highly radioactive wastes from nuclear power stations must be safely disposed of for centuries, something that as yet no one has figured out how to do.

Resource Depletion

Coal, oil, natural gas, and other fossil fuels, the so-called nonrenewable fuels, are not in infinite supply. Take petroleum, for example. The first commercial oil well was drilled in 1859 in Titusville, Pennsylvania. One hundred and thirty-five years later there are more than 600,000 oil wells in over 100 countries. The Washington, D.C.–based Worldwatch Institute estimates that in 25 to 50 years — a mere 200 years after its commercial discovery — virtually all accessible supplies of this crucial resource, which not only fuels our homes and our cars but also is the primary feedstock for countless products from paints to plastics, will be used up.

Ozone Depletion

One of the most troubling environmental impacts of the common air conditioner is its reliance on Freon, a hydrochlorofluorocarbon (HCFC-22), for cooling. By now everybody knows that chlorofluorocarbons (CFCs) have been linked with the depletion of protective ozone in the earth's upper atmosphere. Because HCFCs contain hydrogen, they have a shorter life span and do not cause as much damage to the ozone layer as CFCs. Whereas the U.S. government has called for a phaseout of CFCs by 1996, HCFCs are slated to be phased out by 2030. The phaseout date for HCFCs may be accelerated, however, because recent studies suggest they cause more ozone depletion than was originally believed.

Environmental Products

Whether you're planning a new home or retrofitting an existing one, there are a number of steps you can take to save energy and ease your impact on the planet. These measures run the gamut from plugging air leaks to investing in an efficient new heating system, preferably one that does not rely on fossil fuels. The specific remedies will depend to a great extent on the particular climate you live in. In cool climates, the lion's share of your building or remodeling dollar should go toward buttoning up for winter comfort. If you live in the hot and humid Southeast, ventilation and shading are your paramount concerns. In temperate areas, where summers and winters are either equally harsh or equally mild, your investments in energy conservation should be divided accordingly. In the hot and arid Desert Southwest, proper shading is everything. No matter where you live, your climate can provide clues on how to make your home comfortable and economical to heat and cool. Consider the following common-sense approaches before your think in terms of actual hardware for heating and cooling:

Passive Solar Strategies

Whenever you plan a structure or renovate, there are a variety of design-related considerations to keep in mind.

Solar Orientation — Almost anywhere you are in the United States, you want all the sun you can get. This means that the ideal house will face south, or slightly east of south. A south-facing house will let the most sun in through the windows in winter, when you want it most. Building on southern slopes multiplies the benefits.

Thermal Mass — A commonly used rule of thumb holds that the south-facing window area should not exceed 7 percent of your total floor area unless you have some means of heat storage. The simplest heat storage is a concrete slab, masonry floor, or fireplace that will absorb the heat in sunlight and retain it, re-radiating warmth to your rooms long after the sun has set. Without this thermal mass for heat storage, your house will overheat, even in the middle of winter, and cool off quickly when the sun goes down.

Shading — Large shade trees on the east and west sides of your house supply cooling shade (but don't put them on the south, where even deciduous trees can block vital sun in winter and actually *increase* your heating bills). Properly sized roof overhangs shade windows in summer, when the sun is high in the sky, but don't block the sun's warming rays in winter. Exterior shades, awnings, or shutters can also provide shade for windows that are exposed to direct sunlight.

Windbreaks — Evergreen hedges or shelterbelts buffer a home against cold winter winds, as can an earth-bermed wall, nearby buildings, or hills. Locating garages and porches on the north and west sides of the house, as well as storage areas, stairs, halls, and laundry and other rooms that don't need to be heated to the same exacting standards as living areas, can provide further protection.

Daylighting — Design your rooms with windows and/or skylights for optimum natural light. This will provide natural ventilation and also reduce electrical demand for lighting during the day.

Caulking and Insulation

Invest in an airtight, highly insulated structure. Insulate well beyond the minimum recommendations offered by the federal Department of Energy or your local building department (see "Insulation," pages 126–129). Your local library should have this information, or you can call the Energy Efficiency and Renewable Energy Clearinghouse at (800) 428-2525. The additional insulation can pay for itself swiftly in lower energy bills. Proper caulking and insulation is as important in warm climates to keep heat out as it is in cool climates to keep it in. For specific products, see "Insulation," pages 126–129, and "Caulks and Sealants," pages 130–132. As you tighten up your home, be sure to make as great an effort to reduce sources of indoor pollution and provide for adequate ventilation (see "Mechanical Ventilation," pages 154–157).

Energy-Efficient Windows

High-quality, energy-efficient windows and skylights are another excellent investment. In fact, it makes no sense to build a house with highly insulated walls if all the heat will simply be sucked out through the windows. If you live in a cool climate, look for windows with high R-values (R-value is a measure of the window's ability to keep heat from flowing outdoors). In warm climates, models with low shading coefficients are best (shading coefficient is a measure of a window's ability to transmit solar heat). See "Windows," pages 141–144.

Natural Ventilation

For comfort where summers are hot and humid, especially in the moist heat belt that extends from New York City to New Orleans and Miami, a house must be sited to catch prevailing breezes and designed for ventilation — both cross-ventilation and upward ventilation, which allows warm air to flow up and out the top of the house while cooler air flows in on the ground floor. This means you should bypass sites without breezes. Build in cross-ventilation by placing operable windows on opposite walls

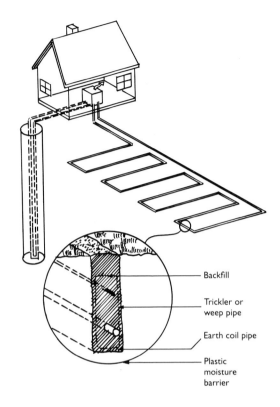

Backfill

Trickler or weep pipe

Earth coil pipe

Plastic moisture barrier

In a ground-source heating system in winter, a nonfreezing liquid circulating through pipes in the ground or in a well absorbs heat from the earth and carries it to a heat pump that extracts the warmth, compresses it to a higher temperature, and distributes it throughout your home. Ground-source systems, which are extremely energy efficient, can also air-condition your home in summer. The newest systems do not use ozone-depleting chemicals and are even more efficient.

in your rooms. Make a cool screened porch one of your main living rooms during the warm months. High ceilings and operable clerestory windows on the leeward side of the house (windows set way up high, where the walls meet the ceiling) together comprise a great system for expelling warm air. For maximum upward ventilation, you may even want to build your house on stilts, like early settlers in Florida and Louisiana did, and put a cupola on top so that warm air can escape.

Ventilating your walls and roof is almost as important as ventilating the inside of your house. If you're reshingling your roof, add ridge vents at the peak and soffit vents at the eaves; these can cut heat flow into the house by up to 35 percent. If you're building new, consider one of the state-of-the-art building technologies such as vent-skin construction, which expels the sun's heat from within the walls before it can reach the interior of the house. Warm air rises through the vent spaces and exits at the roof peak, while cooler replacement air enters through screens at the foot of the vent-skin walls and through soffit vents.

Environmental Heating Systems

Ground-Source Systems

From an environmental point of view, for a replacement heating system, or a primary heating and cooling system for a new house, ground-source systems are tough to beat. No matter where you live, the underground temperature remains relatively constant all year long, even though outdoor temperatures vary widely. Ground-source systems harness this renewable, natural supply of energy and use it for heating and air conditioning. In winter, a nonfreezing liquid circulating through underground pipes absorbs heat from the earth and carries it to a heat pump that extracts the warmth, compresses it to a higher temperature, and distributes it throughout your home. In summer, the unit extracts heat from your house and transfers it back to the circulating liquid in the underground pipe system, where it dissipates in the cooler earth. The pipes are usually laid horizontally three to five feet below the ground in the northern United States and four to six feet deep in the South. They can also be installed horizontally in a nearby pond or vertically in a well. Ground-source heating is not to be confused with geothermal heating, which uses heat directly or indirectly from geothermal wells or springs.

Ground-source systems are extremely energy efficient because they transfer heat rather than produce it. They typically deliver three to four times more energy than they consume. They also provide hot water, eliminating the need for electric or even solar water heaters. These advantages outweigh the systems' major drawback — all models currently use ozone-damaging Freon.

A variety of different systems are available in the United States. Although ground-source heating and cooling is not common, it does have a long track record of good performance. Installation, which includes excavation for the network of pipes, can drive up costs 20 to 25 percent above that of conventional heating, cooling, and hot-water systems. However, ground-source systems pay for themselves quickly in vastly reduced fuel costs, and some utilities offer rebates to homeowners who install them.

Hydronic Systems

While ground-source systems tap the most environmentally benign form of energy, hydronic (hot water) systems are the best way to distribute the heat throughout your house. The best hydronic systems are radiant floors, in which tubing is installed in the floor to distribute hot water — making the whole floor, in effect, a radiator — and baseboards, in which the hot-water tubing runs around the perimeter of the room. Hydronic heating is superior to hot-air systems in several respects: although the initial investment is higher, your money is paid back in energy savings because radiant energy, unlike hot air from forced-air systems, doesn't collect at the ceiling and so you can set your thermostats lower.

Hydronic heating is also better for indoor air quality. Unlike air systems, they have no network of ducts that can channel airborne pollutants from one part of the house to another, and create no air currents that blow dust and other pollutants off flooring, carpeting, and furnishings. Anyone who has experienced warm floors in winter will also attest to the fact that these are the most comfortable heating systems around. However, do avoid baseboard radiators that incorporate plastic tubing and/or plastic radiator covers as they are a potential source of polyvinyl chloride and other air pollutants. Instead, select hot-water baseboards with copper tubing and metal covers.

Woodstoves and Fireplaces

Since the 1970s when hundreds of thousands of energy-conscious Americans installed wood-burning stoves in their homes — and soon discovered that they were blanketing the atmosphere with noxious pollutants — a new generation of stoves that burn cleanly and much more efficiently has emerged. Nevertheless, it is not wise to make one of these your primary heat source because burning wood simply is not an efficient use of this increasingly scarce resource. If wood burning is to be an adjunct heat supplier in your home, however, the following are good choices.

Iron Stoves — When state and federal regulators cracked down on air pollution from wood-burning stoves, some manufacturers added catalytic converters to their models. Others achieved a cleaner burn by employing sophisticated airflow designs. Still others have turned to a new form of wood fuel — pellets. Because pellet stoves burn cleanly and with great efficiency, and in addition make use of waste wood that has been pulverized and then compressed, they are the environmental first choice.

A circa-1975 woodstove probably burned at an overall efficiency of less than 60 percent, meaning that less than 60 percent of the potential heat in the wood actually ended up in your room. The best pellet-burning stoves now operate at efficiencies of almost 80 percent. When shopping for a woodstove, look for models with the best emissions ratings and those that burn most efficiently. Suppliers of the top pellet burners according to emissions ratings by the U.S. Environmental Protection Agency are listed on page 152.

Masonry Heaters — These Eurostyle stoves have been used for centuries in Europe and are now attracting an ever-larger following on this continent. Unlike traditional North American stoves, which are made of iron, these stoves are made of masonry; unlike iron stoves, masonry stoves are heat-storage systems. Designs differ, but the operative principle remains the same: heat built up rapidly in the firebox travels through a long maze of built-in channels, or flues, is stored in the stove's masonry mass, and is released into the room over a period of hours. Consequently, masonry heaters require wood only once or twice a day, and they burn at efficiencies of 70 to 90 percent. They also burn cleanly, don't get hot to the touch, and come in a variety of styles, including models with cooktop and bake oven. They can be faced with brick, stucco, soapstone, or glazed tile. They're a handsome architectural feature and form of thermal mass that will last as long as your house.

Not surprisingly, masonry heaters are also expensive. And they work best in well-insulated homes with open floor plans that allow the heat to radiate freely.

Efficient Fireplaces — In the conventional fireplace, 90 percent of the heat escapes up the chimney. Such fireplaces are also dirty, generating extremely high rates of particulates, tiny bits of unburned hydrocarbons that can lodge in the lungs and cause respiratory problems. If you can't live without the crackle and pungent aroma of a fireplace, install a Rumford fireplace. These traditional eighteenth-century designs are shallow to radiate heat better and have a streamlined throat to carry away the smoke with less heat loss. To install a Rumford fireplace in your home, you'll need a specially designed throat, cast-iron dampers, and smoke chambers. These are still made by a handful of companies.

One final word about wood-burning stoves and fireplaces: be sure to have fresh air intakes that bring in outside air for combustion. In today's tightly constructed homes, this is especially important to keep noxious combustion fumes from "backdrafting" into the house. Monitor the chimney to avoid chimney fires, choose nonflammable roofing material, and keep your smoke detectors in good working order.

Alternative Cooling Systems

Alas, there is still no such thing as an environmentally sound air conditioner. Although some models are much more energy efficient than others, it is difficult to wholeheartedly endorse them because they all currently use the hydrochlorofluorocarbon Freon as the coolant. Although a few air conditioners for commercial and industrial use that rely on alternative coolants are available, there is not yet a home model that does not make use of ozone-depleting chemicals. Until manufacturers resolve this issue, the most prudent course of action is to design or remodel your home to make the most of shading and natural ventilation. Ground-source heating and cooling systems are energy-efficient alternatives for new homes or those in which both the cooling and heating system needs to be replaced, although these, too, use Freon. There are a few other alternatives to the conventional air conditioner that are worth your consideration.

Ice and/or Cold-Water Storage Systems — A recent development in home cooling, as well as space and hot-water heating, these systems store ice and heat in two large tanks: one pressurized for domestic hot water and the other nonpressurized for space heating or cooling. The storage tanks are what make them different from conventional heat pumps, by enabling the compressor to operate only at night when electric rates are low. This not only saves money, shortening the time it takes for the system to pay for itself in energy savings, but also reduces your electric utility's need to build additional generating capacity to satisfy power demands during the day, when they are driven sky high by ordinary air conditioners. On summer nights, the heat pump extracts warmth from the water in the space-conditioning storage tanks, creating cold water or ice with which to cool the home the next day. The extracted heat is transferred to the domestic hot-water tank. During the colder months, the heat pump works in reverse and the water is heated to warm your rooms.

Because ice and/or cold-water storage systems cost 25 to 40 percent more than conventional air conditioning, they make sense financially only in hot climates where air conditioning loads are high and electric rates are steep. These systems are more cost competitive if your electric utility offers special off-peak rates.

Indirect Evaporative Air Cooling — Evaporative cooling is a time-tested method employed by "swamp coolers" and other traditional systems in arid areas. They work by passing air through a wet medium; as some of the moisture evaporates, the air temperature drops. An indirect evaporative cooler produces cooled air via a two-step process. In the first stage, warm, dry air is passed over cooling fins that contain water. Because the air does not contact the water in the fins, the air is cooled without adding moisture to it. This pre-cooled air is then drawn into an evaporative cooler, where its temperature is lowered even further. Such units require as much as 80 percent less electricity than conventional air conditioners to lower a room's air temperature by twenty degrees. Indirect evaporative air coolers are cost competitive with conventional air conditioning in hot, arid areas.

Variable Air Volume Systems — Unlike conventional air conditioners, which remain at a pre-set cooling level and air flow, these energy-conserving designs automatically adjust to changing temperatures. Although they are energy-efficient, they still make use of ozone-depleting Freon.

Suppliers

GROUND-SOURCE HEATING AND COOLING SYSTEMS

Command Aire, Geothermal Heating Systems
P.O. Box 2015, Lower Burrough, PA 15068; (412) 335-3303

Delta T Corporation
10520 Route 6N, Albion, PA 16401; (814) 756-5848

Earth Systems International, Inc.
258 McBrine Drive, Kitchener, Ontario, Canada N2R 1H8; (800) GO-EARTH

GSDX (Ground Source Direct Exchange), **U.S. Power Climate Control, Inc.**
881 Marcon Boulevard, Allentown, PA 18103; (800) 669-1138 or (215) 266-9500

Tetco (Thermal Energy Transfer Corporation)
1290 U.S. 42 North, Delaware, OH 43015; (800) 468-3826

WaterFurnace International, Inc.
9000 Conservation Way, Fort Wayne, IN 46809; (219) 478-5667

HYDRONIC HEATING
Radiant Floor Systems

Aztec Intl.
2417 Aztec Road N.E., Albuquerque, NM 87107; (505) 884-1818

EasyHeat, Inc.
31977 U.S. 20 East, New Carlisle, IN 46552; (219) 654-3144

ESWA STK Canada
5132 Timberlea Boulevard, Mississauga, Ontario, Canada L4W 2S5; (416) 624-2723

Flexwat Corp.
2380 Cranberry Highway, West Wareham, MA 02576; (508) 291-2000

Heatway
3131 W. Chestnut Expressway, Springfield, MO 65802; (417) 864-6108

Infloor Heating Systems
920 Hamel Road, P.O. Box 253, Hamel, MN 55340; (612) 478-6477

Wirsbo Company
5925 148th Street West, Apple Valley, MN 55124; (612) 891-2000

Baseboard Systems

Runtal Radiators, Runtal North America, Inc.
187 Neck Road, Ward Hill, MA 01835; (800) 526-2621

WOOD-PELLET-BURNING STOVES

Emerald 2000, Guertin Brothers Industries, Inc.
18931 59th Avenue, NE #4, Arlington, WA 98223; (206) 653-5505

Model Dynasty, Winrich International
8601 200th Avenue, Bristol, WI 53104; (414) 857-7800

Whitfield Advantage WP-2, Pyro Industries, Inc.
695 Pease Road, Burlington, WA 98233; (206) 757-9728

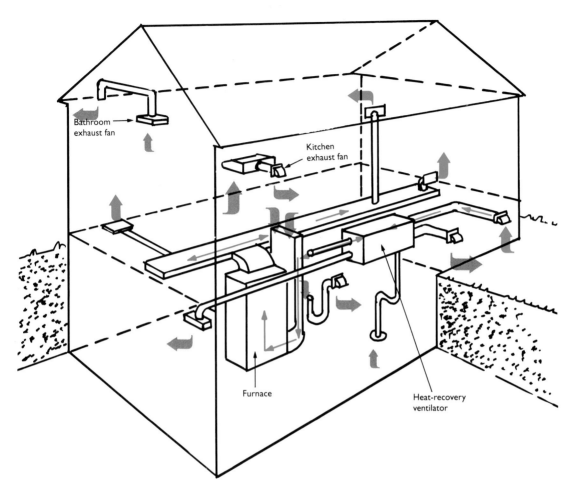

Bathroom
exhaust fan

Kitchen
exhaust fan

Furnace

Heat-recovery
ventilator

A heat-recovery ventilator (HRV) is an important part of a comprehensive strategy to rid a home of pollutants. It provides continuous ventilation by transporting stale, polluted air (shown in gray) from your rooms or the return air trunk of your furnace to the HRV, which uses the heat from this warm air to heat cold, fresh incoming air (shown in color). Additional exhaust capacity is provided by fans in the kitchen and bathroom.

MASONRY HEATERS

Biofire, Inc.
3220 Melbourne, Salt Lake City, UT 84106; (801) 486-0266

Royal Crown European Fireplaces, Inc.
333 East State Street, Suite 206, Rockford, IL 61104; (800) 373-2042 or (815) 968-2022

TuliKivi Swedish Soapstone Heaters
P.O. Box 300, Schuyler, VA 22969; (800) 843-3473 or (804) 831-2228

RUMFORD FIREPLACES

David Lyle
P.O. Box 300, Acworth, NH 03601; (603) 835-2318

Superior Clay Corp.
P.O. Box 352, Uhrichsville, OH 44683; (800) 848-6166

ICE STORAGE AND COLD-WATER STORAGE SYSTEMS

Baltimore Aircoil Company
P.O. Box 7322, Baltimore, MD 21227; (410) 799-6200

THP/3, Phenix Heat Pump Systems
8390 Gerber Road, Elk Grove, CA 95828; (916) 689-8111

INDIRECT EVAPORATIVE AIR COOLERS

DePeri Manufacturing
P.O. Box 280815, North Ridge, CA 91328; (818) 885-0011

Vari-Cool
P.O. Box 548, Hastings, NE 68902; (402) 463-9821

VARIABLE AIR VOLUME COOLING SYSTEMS

Harmony II Zoning System, Lennox Industries
P.O. Box 799900, Dallas, TX 75379; (214) 497-5000

MECHANICAL VENTILATION

In an energy-efficient house, ventilation is critical. Modern construction techniques, such as lining the entire house with an air infiltration barrier, can seal a house up so tight that the ventilation rate is 0.2 air changes per hour or even lower. This means that one complete house volume of fresh air is exchanged every five hours at most. In an unventilated house that tight, fumes from heating, cooking, solvents, cleaners, and smoking, as well as moisture from breathing and bathing, can make the air very unhealthy.

Conventional Products

The average American home may have an exhaust fan in the bathroom, which can be turned on after showering or bathing to expel overly moist air. It may also have a range hood, which may or may not be vented to the outside to expel combustion pollutants. For continual ventilation, most houses rely on air leaks in the building shell. Older homes in the United States, even those that have been fairly extensively weatherized, are typically leaky enough to undergo one air change every hour or two. However, relying on air leaks is rather a hit-or-miss approach because the air-exchange rate — and therefore the quality of the air you breathe — depends on such variables as how windy it is outside, the location of the leaks, and how many times you open your door every day, factors that are mostly out of your control. Even periodically opening your windows may not be sufficient, and it certainly isn't energy efficient. Only continual, controlled ventilation can guarantee that the air inside your home is healthy.

Health and Environmental Concerns

A typical house harbors dozens of polluting products. One study sponsored by the Envi-

FIVE STEPS TO HEALTHY INDOOR AIR

From design through construction, remodeling, and maintenance, there are a number of steps you should take to insure that the air inside your house is healthy.

1. Start by identifying the main sources of pollutants in your home and take steps to remove them or reduce the pollution they emit (see "A Guide to Home Pollutants," pages 245–249). If you're designing a new home, avoid using products most likely to pollute your household air.

2. Test radon levels in your house, and take measures to reduce them as necessary (see "Radon Prevention Systems," pages 81–90).

3. Whether you're remodeling or building new, make the most of natural ventilation in your home (see "Heating and Cooling," pages 145–154).

4. Install a heat-recovery ventilator to expel stale, polluted air without wasting energy.

5. For extra protection, use an air filter and purification system that can cleanse the air of both polluting particles and gases (see "Air Filtration and Purification," pages 157–159).

---

ronmental Protection Agency found more than 350 volatile organic chemicals (synthetic, petroleum-based chemicals that volatilize, or become a gas, at room temperature) in the air of a single house. Many common household products, from cleaners to furniture polish, are culprits. So are common construction and decorating materials: wallpapers, paints, upholstered furniture, even permanent-press bedding. Particleboard and some construction adhesives are among the worst offenders because they release formaldehyde into the air. Combustion gases from the burning of heating or cooking fuels in the house, most particularly nitrogen dioxide and carbon monoxide, can cause both acute and chronic health problems if they aren't adequately vented. Animal dander, dust, dust mites, mold and mildew, and other allergens are in the air in every house. What is more, naturally occurring gases such as radon can be extremely hazardous to your family's health (see "A Guide to Home Pollutants," pages 245–249).

Environmental Products

An important part of any overall strategy to rid a home of these pollutants is to insure that the house constantly "breathes" in fresh air and expels stale, polluted air. At this time, the best way to accomplish this, without wasting energy expended to heat and cool your house, is with a heat-recovery ventilator. Also known as an air-to-air heat exchanger, it uses the heat from warm, stale air being exhausted from the house to heat the cold, fresh incoming air. By the time the fresh air enters your house, it can be quite a bit warmer than the outdoor air. These devices can also help ventilate an air-conditioned house in the summer, cooling incoming fresh air instead of warming it. Even if you have a heat-recovery ventilator, you still need a properly sized exhaust fan in the kitchen to expel pollutants when you are cooking.

Most heat-recovery ventilators are intended to ventilate the whole house and require their own ductwork. The units typically are smaller than the air handler of a central home heating and air-conditioning system. Fresh air is drawn in an intake and passes through the heat-recovery ventilator, where it is warmed by as much as 70 percent of the heat in stale air on its way outdoors. Different models use different heat-transfer mechanisms; ideally, the two air streams do not mix, although in some units there is some cross leakage. The warmed fresh air is then routed to the rooms where you spend most of your time, particularly the bedrooms, playrooms, and nursery. The return, which carries moist and polluted air back to the heat-recovery ventilator, is usually located in the bathroom in place of the conventional bathroom ceiling fan. The heat-recovery ventilator normally is on around the clock, breathing for your home. You can manually raise or lower the air-exchange rate as necessary; for example, you might want to turn it up if you are having a party, or hook it up to a timer so that it comes on once a day or so while you are on vacation. Whole-house heat-recovery ventilators cost approximately $700; $3,000 total with ductwork and labor. Consequently, it makes most sense to install them when you're building a new home, adding a central heating and cooling system, or doing other extensive renovation.

There are some models designed to ventilate only a room or two. They mount in the window or wall like a small room-sized air conditioner, so they're the easiest to add to an existing structure or apartment.

Suppliers

Airxchange, Inc.
401 VFW Drive, Rockland, MA 02370; (617) 871-4816

American Aldes Ventilation Corp.
4537 Northgate Court, Sarasota, FL 34234; (813) 351-3441

Bard Manufacturing Company
1914 Randolph Drive, Bryan, OH 43506; (419) 636-1194

ClimateMaster
P.O. Box 25788, Oklahoma City, OK 73125; (405) 745-6000

Comfort Ventilator, United Technology Carrier
7310 W. Morris Street, P.O. Box 70, Indianapolis, IN 46206; (317) 243-0851

Command-Aire
P.O. Box 7916, Waco, TX 76714; (817) 840-3244

Environmental Air Ltd.
P.O. Box 10, Cocagne, NB, Canada E0A 1K0; (506) 576-6672

E-Z Aire and Vent II, DesChamps Laboratories, Inc.
P.O. Box 440, 17 Farinella Drive, East Hanover, NJ 07936; (201) 994-4660

Honeywell, Inc.
Residential and Building Controls Division, 1985 Douglas Drive North, Golden Valley, MN 55422; (612) 542-3357

Perfect Aire, Research Products Corp.
P.O. Box 1467, Madison, WI 53701; (608) 257-8801

Thermax, Kooltronic, Inc.
57 Hamilton Avenue, P.O. Box 300, Hopewell, NJ 08525-0300; (800) 929-0682 or (609) 466-8800

VanEE
P.O. Box 10416, Minneapolis, MN 55458; (800) 667-3717

Venmar Ventilation
1715 Haggerty Street, Drummondville, Quebec, Canada J2C 5P7; (819) 477-6226

Vent-Aire, Engineering Development, Inc.
4850 Northpark Drive, Colorado Springs, CO 80918; (719) 599-9080

In addition to replacing stale air with fresh air, a home must be capable of cleaning its air to keep levels of indoor pollutants low. This can be accomplished with a home air filtration and purification unit.

A good air-cleansing system works on two fronts: it filters out particles, small solid or liquid substances suspended in the air throughout the house; these invisible particles, including dust, aerosols, and tobacco smoke, are of concern because they can be inhaled deeply into the lungs, with potential health effects ranging from lung irritation to allergies and asthma to cancer. It also removes gaseous pollutants, both the by-products of combustion in your stove,

A good air-cleansing system will keep levels of indoor pollutants low. The most advanced air filters involve a three-step process. A pre-filter removes large particles like pollen; a HEPA (high-efficiency particulate absolute) filter traps the most minute particles, such as those in tobacco smoke; and a third medium, usually carbon, works on formaldehyde and other gaseous pollutants.

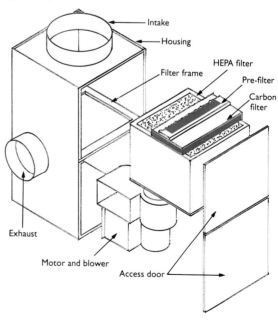

Intake
Housing
HEPA filter
Filter frame
Pre-filter
Carbon filter
Exhaust
Motor and blower
Access door

oven, fireplace, and heating system, and the volatile organic compounds that are released by countless petrochemical products used to build, decorate, and maintain the typical American house (see "A Guide to Home Pollutants," pages 245–249). Although the U.S. Environmental Protection Agency has not taken a position either for or against the use of these devices, they are an important component of any overall strategy to maintain high-quality indoor air and safeguard your family's health.

Conventional Products

In the average house, air cleansing begins and ends with the range hood in the kitchen. The typical nonvented or recirculating range hood doesn't do you much good at all. The best models may remove grease and smoke by forcing cooktop air through a filter before sending it back in the room, but they don't remove other types of particles and gaseous pollutants.

When choosing an air filtration and purification system for your home, you should look for the following: it should remove a high percentage of the particles in the air — that is, it should be highly efficient; it should be capable of removing both particulate and gaseous pollutants; and it should be easy to maintain and not suffer appreciably in performance between maintenance periods. Some of the more common air-cleaning devices fail the above tests. For example, the typical ion generator neither filters nor purifies; it charges particles in the air, causing many of them to fall to the floor, where levels can build up and pose even more of a threat to infants and children. Electrostatic precipitators also work by giving particles a negative charge; the particles are then attracted to a positively charged plate, grid, or screen. As more and more particles are trapped, the unit's efficiency suffers. Every three or four weeks the plate, grid, or screen must be removed and cleaned with a solvent-based degreaser. And

these systems do nothing about gaseous pollutants. They also produce ozone gas, a respiratory health hazard.

Environmental Products

A much more advanced system of air cleaning is now available. Typical units are capable of extremely high efficiencies of up to 99.9 percent and can remove particles as minute as 0.3 microns (1 micron equals $\frac{1}{25,400}$ inch; the average thickness of a human hair is 100 microns, that of a typical grain of ragweed pollen, 20 microns). They also remove gaseous pollutants. Most units involve a three-step process: a prefilter or series of them remove relatively large particles, such as pollen and dander. A HEPA (high-efficiency particulate absolute) filter effectively traps the most minute particles, including those in tobacco smoke and dust. Such filters consist of a bed of randomly positioned glass fibers. As the particles collide with the fibers and adhere to them, the passages become smaller and the filter becomes even more efficient. HEPA filters are so effective, in fact, that they are used in operating rooms, pharmaceutical and computer-chip manufacturing plants, and other places where the highest air quality is required. Because the HEPA filter removes only airborne particulates, a third component works on formaldehyde and other volatile organic compounds commonly found in houses, as well as combustion gases. The filtering medium is usually activated carbon, the same medium used in many domestic water filtration systems. Other media, including purafil (alumina impregnated with potassium permanganate, often used to filter out odors) and zeolite (a pumice-like volcanic substance especially effective at removing urine odors in bathrooms and around litterboxes) are also used, and a combination of media can be special ordered.

These combined air filtration and purification units can be integrated with your home's central air system; they're usually mounted next to the system's air handler. The better models have their own fans built in. They can be manually controlled or automatically controlled via a programmable computer chip. Unlike heat-recovery ventilators, air filters generally are not on around the clock. You can turn them on at night, when allergies are acting up, when you are having a party, or when you're cooking or engaged in some special activity that is apt to generate pollutants. Portable units are available for use in homes without central air systems. However, whole-house units, while more expensive, are quieter and, of course, cleanse the air in all your rooms.

Both the HEPA filter and the activated carbon need to be replaced at least once a year; the more expensive models have automatic monitors that will warn you when it's time to replace them. Both the HEPA and the carbon filters are easily accessible and simple to replace.

Suppliers

Aireox
P.O. Box 8523, Riverside, CA 92515; (714) 689-2781

Airstar-5, Allermed Corporation
31 Steel Road, Wylie, TX 75098; (214) 422-4898

CRSI
P.O. Box 418, Plainfield, IN 46168; (317) 839-9135

Dust Free
P.O. Box 519, Royse City, TN 75189; (800) 441-1107

E. L. Foust
P.O. Box 105, Elmhurst, IL 60125; (800) 225-9549

Euroclean
907 West Irving Park Road, Itasca, IL 60143; (800) 545-4372

FILTRx
11 Hanson Avenue, New York City, NY 10956; (914) 638-9708

Healthmate and Healthmate Plus, Austin Air, Environmental Construction Outfitters
44 Crosby Street, New York, NY 10012; (800) 238-5008 or (212) 334-9659

King Aire
P.O. Box 398, Noblesville, IN 46060; (800) 991-5464

Nigra Enterprises
5699 Kanan Road, Agoura, CA 91301; (818) 889-6877

Thurmond Air Quality Systems
P.O. Box 940001, Plano, TX 75049-0001; (800) AIRPURE or (241) 422-4000

Compared to water and gas, electricity is the youngest service, and the one that truly makes the modern home possible. Electricity had no purpose in houses until it became a means to light them. This change was sparked by the development of a practical incandescent bulb in 1879 by Thomas Edison. Edison, also something of a businessman, realized that the market for his bulbs would be soft if no one had access to the form of energy that made them work. By 1882 he was operating the Pearl Street Station in New York City, the first power plant specifically designed to supply electricity on demand to consumers.

In electricity's short one hundred years, we've become so dependent on it that we can hardly envision the world without it. Imagine your house without electric lights, the stereo, the VCR, the hair dryer, and the food processor and you begin to understand the way this relatively new technology has shaped our lives.

Conventional Products

Electricity typically is generated in large, centralized generating stations. In these power plants, coal, oil, natural gas, or nuclear fuel are used to heat water until it becomes steam, which turns the turbines that produce the electricity. At hydropower facilities, the turbines harness the energy embodied in water spilling over a dam. Electric power stations are linked to one another and to your home by an extensive grid of overhead and underground transmission lines.

Health and Environmental Concerns

Electricity is often touted as the cleanest form of energy. This may be true once it reaches your home, but it is certainly not the case where the energy is produced. Coal mining is dangerous and disturbs the land. The supply of imported oil is subject to politically motivated disruptions, as we learned all too well during the 1970s. Burning coal and oil generates air pollution, including greenhouse gases such as carbon dioxide, which have been linked to global warming. The hazardous radioactive wastes produced at nuclear power plants pose serious, long-term problems. Hydroelectric projects are notorious for destroying the environments of rivers and lakes with their huge dams and reservoirs. And, with the exception of hydropower, all forms of electricity generation consume nonrenewable resources.

Environmental Products

Sunlight and electricity are both forms of energy. Photovoltaics (PV) is the technology of converting sunlight into electricity. Once in place, a PV system requires no machinery, no moving parts, and no fuel, other than sunshine. Contrary to popular belief, photovoltaics is not the wave of the future — it's available right now.

In 1954, Bell Laboratories achieved the breakthrough that brought photovoltaics out of the laboratory and into the realm of practical application. Less than three years after Bell Labs announced its new PV cell, the space race between the U.S. and the Soviet Union began in earnest, and the technology was boosted by the rush to find a way to power orbiting spacecraft. In those days, photovoltaics was prohibitively expensive — about $2,000 per watt. The cost has since plummeted to $5 per watt, making solar electricity economical for a good number of homeowners. Prices continue to come down, making the residential roof the power plant of choice. In fact, there are approximately 100,000 photovoltaic houses in this country. Photovoltaics has come so far that it is worth your time to review the costs and benefits of a solar electricity system for your home.

EMF-SHIELDED WIRING AND APPLIANCES

Wherever there is electricity, there are electromagnetic fields. EMFs are emitted by all wires and appliances that carry current — in other words, all electrical devices that don't run on batteries. For more than twenty years, scientists have debated whether electromagnetic fields represent a threat to human health, never coming even close to a consensus, although recent research in Sweden shows the strongest connection yet between EMFs and an increase in leukemia among both children and adults (see "A Guide to Home Pollutants," pages 245–249).

You don't have to wait until scientists resolve the EMF issue in order to protect yourself and your family from potentially adverse health effects. BX electrical cable, which is encased in metal (as apposed to Romex, which is sheathed in plastic) effectively prevents the wiring from emitting electromagnetic fields. BX is most likely to be present in new homes and in houses in the city; Romex in older homes and non-urban areas. If your home has the plastic cable, you can replace it with BX; this makes especially good sense when you're doing extensive renovation.

Circuit-breaker panels are easily shielded with a sheet of $\frac{1}{2}$-inch steel or lead. Computer and TV screens, big concerns for parents and most office workers, can be fitted with a screen-over-the-screen that not only protects against EMFs but also reduces glare, making the screens easier on the eyes. A few EMF-shielded lighting fixtures are also available (see "Lighting," pages 227–232). For other electrical appliances, the best course of action is to stay well away from them when they are in operation.

Suppliers

EcoVision, Blackhawk Computers
38 Main Street, Chatham, NY 12037; (518) 392-7007
Low-radiation computer monitors.

Fairfield Engineering
P.O. Box 139, Fairfield, IA 52556; (515) 472-5551
Monitor shield.

NoRad Corporation
1549 11th Street, Santa Monica, CA 90401; (800) 262-3260 or (213) 395-0800
Ultra Glass and Radiation Shield EMF-shielding devices for computer and TV screens.

Safe Technology Corp.
1950 Northeast 208 Terrace, Miami, FL 33179; (800) 638-9121
Reduced-radiation monitors.

Sigma Designs
46515 Landing Parkway, Freemont, CA 94538; (510) 770-0100
Low-radiation monitors.

The workhorse of a photovoltaic system is the photovoltaic cell. The heart of the photovoltaic cell is silicon, one of the most common materials on earth. There are three types of cells: crystalline, polycrystalline, and amorphous. Crystalline cells are made of slices from one large cultured silicon crystal. They are the most efficient and usually the most expensive. Polycrystalline cells are slightly less efficient than single crystalline cells, but also less expensive. Amorphous cells are noncrystalline; thin layers of silicon are applied to a base in a process not unlike lamination. These cells are less efficient than their crystalline counterparts, but they also cost less. When sunlight strikes a cell, light energy, or photons, knocks loose electrons, creating direct current. The electrons are collected by metal contacts arranged into a small grid.

To increase power, cells are wired to other cells. Twenty or more cells packaged under glass or plastic are known as a module. Groups of modules are wired together to form an array.

It's important to keep in mind that PV cells produce 12-volt direct current (VDC), not the standard 110-volt alternating current (VAC). Most household appliances are available in VDC. If you want to keep some of your old appliances, you can buy an inverter to convert the 12VDC to 110VAC. Depending on the type of system you choose, you may also need batteries to store the power for nights or rainy days, a backup generator, a voltage regulator to control the flow of current to the batteries, and some monitoring gauges.

The main expense in PV systems is the number of modules required. With a few lifestyle changes, you can save enormous amounts of electricity and lower the cost of the system. For example, you can switch from incandescent bulbs to compact fluorescents (a good idea in any case) and do without electric resistance appliances such as electric stoves and ovens (buy propane or gas versions), electric coffee perco-

lators (use manual drip filters), and electric blankets (use a down comforter).

Your PV supplier can help you determine whether PV makes sense for you and how big an array your home requires. One way to calculate how many panels you'll need involves using your current electrical usage as a basis for the PV system you plan to install. If you spend $100 per month for electricity that costs $.08 per kilowatt hour, you're using roughly 41,600 watt-hours per day. Assuming a good six hours of direct sunlight each day, you'll need to "collect" about 6,900 watts from the PV array. This equals 115 panels each rated at 60 watts — or 138 after 115 is multiplied by 1.2 to compensate for dust and bird droppings on the panels and other variables. With an average cost of $300 per 60-watt panel you'll spend $41,400 to match your current power needs — which is why it's a good idea to convert to more efficient appliances.

Batteries are the second-biggest expense in a PV system. Unfortunately, lead or nicad (nickel and cadmium) batteries not only add to the cost of a PV system but are also toxic and must be properly disposed of.

Until recently, most home photovoltaic units were stand-alone systems used in homes in remote areas where it is extremely expensive to tie into existing utility lines — a mountain or desert cabin, for example (see "A Weekend Retreat," pages 53–59). A basic model is a 48-watt panel with battery storage that allows you to have two or three fluorescent lights and a television. The retail price of a stand-alone system begins at $1,000 and can go considerably higher, depending on the degree of comfort and convenience you expect.

In increasing numbers, homeowners are investing in grid-integrated systems; in other words, their homes are still hooked up to the local electric utility. No batteries are required, decreasing costs and eliminating maintenance and disposal problems. At night or during inclement weather, electricity is brought in from the grid to power the home. When the sun shines and the PV array's output exceeds demand, electricity is sold back to the utility.

Most people assume that solar energy is only feasible in the sunniest areas of the country. This isn't necessarily the case. Whether or not photovoltaics is cost effective for you depends not only on sunshine, or insolation, levels in your area but also the cost of electricity. New Englanders, for example, get only about half as much sunshine as residents of the sunniest regions. They therefore need larger — and more expensive — arrays. However, electricity rates in New England are also relatively high, so the potential savings in monthly utility bills is higher as well. This means that the systems will pay for themselves relatively quickly. Only in areas where low utility rates are coupled with low levels of sunshine — the Pacific Northwest, for instance, where homeowners have access to cheap hydroelectric power — is photovoltaics an economically unattractive proposition.

Steven Strong of Solar Design Associates in Harvard, Massachusetts, suggests the following simple way to calculate how close the technology is to the break-even point in your area: just take the projected annual output, in kilowatt hours, of the system you are contemplating, multiply it by a design life of twenty-five years, and divide by the overall system cost, making the necessary adjustments for state tax credits and/or utility rebates, if any, and interest on a loan. Then compare this figure with the electric rates in your area.

PV arrays look a lot better when they're mounted flush with your roof line, provided that the house is oriented toward solar south and the angle of the roof is right (approximately your latitude plus or minus ten degrees). Mounting the panels at a different angle disturbs the silhouette of the roof. The arrays can also be mounted on solar trackers on the ground.

Suppliers

The following companies are distributors or dealers for photovoltaic equipment:

Advanced Photovoltaic Systems, Inc.
195 Clarkesville Road, Lawrenceville, NJ 08648; (609) 275-5000

Atlantic Solar Products
9351J Philadelphia Road, Baltimore, MD 21237; (301) 686-2500

B & E Energy Systems, Inc.
3530 Franklin Road, Bloomfield Hills, MI 48013; (313) 540-9617

Energy Conservation Services of North Florida
619D South Main, Gainesville, FL 32601; (904) 373-3220

Fowler Solar Electric, Inc.
131 Bashan Hill Road, Worthington, MA 01098; (413) 238-5974

Inter-Island Supply
345 N. Nimitz Highway, Honolulu, HI 96817; (808) 523-0711

William Lamb Corporation
10615 Chandler Boulevard, North Hollywood, CA 91601; (818) 980-6248

Photocomm
7735 East Redfield Road, Scottsdale, AZ 85260; (800) 544-6466

Photron, Inc.
1220 Blosser Lane, P.O. Box 578, Willits, CA 95490; (707) 459-3211

Remote Power, Inc.
649 Remington Street, Fort Collins, CO
80524; (303) 482-9507

RMS Electric, Inc.
2560 28th Street, Boulder, CO 80301; (303)
444-5909

Skyline Engineering, Inc.
P.O. Box 134, Temple, NH 03084; (603) 878-
1600

Solar Design Associates
Still River Village, Harvard, MA 01467-0143;
(508) 456-6855

Solar Electric of Santa Barbara
232 Anacapa Street, Santa Barbara, CA
93101; (805) 963-9667

Solar Electric Systems
1244 Bell Avenue, Fort Pierce, FL 34982;
(407) 464-2663

Solar Electric Systems, Inc.
2700 Espanola NE, Albuquerque, NM 87110;
(505) 888-1370

Solec International, Inc.
12533 Chanron Avenue, Hawthorne, CA
90250; (213) 970-0065

Solar Works
64 Main Street, Montpelier, VT 05602; (802)
223-7804

The best source of information and supplies for
do-it-yourselfers is Real Goods, 966 Mazzoni
Street, Ukiah, CA 95482-3471; (800) 762-
7325. The mail-order company offers several
different-size PV packages, in addition to the
Alternative Energy Sourcebook, a 528-page
compendium of information on designing and
building PV and other home solar systems.

ENERGY MANAGEMENT SYSTEMS

By now it should be obvious that the home
heating, ventilation, and air-conditioning
(HVAC) system is becoming much more complicated than the old-fashioned furnace or
boiler in the basement. That's not surprising,
given that the goal of the twenty-first-century,
or even the 1990s, HVAC system is not only to
increase human comfort but also to keep energy
consumption as low as possible; minimize the
depletion of natural resources, including fossil
fuels; reduce global environmental pollution;
and safeguard the quality of air indoors. Indeed, the state-of-the-art HVAC system includes a number of different components:
heating, cooling, heat-recovery ventilation, and
air filtration and purification. The easiest —
and most energy-efficient — way to deal with
this complicated network is with one of the new
programmable home management systems.

Home management devices range from the
simple to the high-tech. One of the simplest is
the basic seven-day programmable system. It
enables you to divide your home into as many
as fifteen to twenty separate rooms or "zones"
and to manually control, at all times, the
amount of heating and cooling in each zone. If
you're watching evening TV in the living room,
say, you can turn down the heat in the rest of
the house and save energy. If you'll be out of
town for five or six days, you can keep the heat
low the whole time and program the unit to
turn it up just before you get back.

The most sophisticated models are tied into a
central home computer. You get room-by-room
control of temperature, humidity, ventilation,
and air filtration. You can program the computer to turn on or shut off any electrical appliance: for example, your coffeemaker can begin
brewing automatically at 6:30 A.M. on weekdays, just as your stereo comes on to wake you
up. Your porch light can be programmed to
turn on when someone approaches the front

door after dark. During vacations, lights and music can be set to go on and off at different times every day to make it seem as if someone is home. These high-tech systems also manage security in as many as two dozen different zones of the home. You can assign separate passcodes for your babysitter or repair person, limiting their access to specific times, days, and even security zones. All this is coordinated at a wall-mounted computer screen that displays your home's exact floor plan and step-by-step instructions and is activated by touch.

Home management systems cost anywhere from $2,000 to $50,000. They can decrease your monthly energy bills 35 to 55 percent.

Which system is best for you? That depends on your lifestyle and budget, the size of your home, the complexity of its systems and how easy or difficult they are to use, and the availability of technical support and repair people in your area. It makes good sense to have at least the basic seven-day programmable system. Without all the bells and whistles — music, coffee-making, and the like — the basic system offers benefits any environmental house should have:

- energy-conserving control of heating and cooling systems (eliminating thermostats)
- indoor air-quality monitoring
- enhanced security
- energy-conserving lighting control

The computer manages these basic tasks very efficiently and cost effectively. And it is easy to use.

Suppliers

Carrier Corp.
7310 W. Morris Street, P.O. Box 70, Indianapolis, IN 46206; (317) 243-0851

Home Automation, Inc.
2313 Metairie Road, P.O. Box 9310, Metairie, LA 70055-9310; (504) 833-7256

Home Manager, Unity Systems, Inc.
2606 Spring Street, Redwood City, CA 94063; (800) 85-UNITY or (415) 369-3233

Honeywell, Inc.
Honeywell Plaza, Minneapolis, MN 55408; (800) 345-6770

INTERIOR SURFACES

Gypsum wallboard is known by many other names, including sheetrock, drywall, wallboard, and plasterboard. No matter what you call it, gypsum wallboard is the most common wall and ceiling material in American homes today. It replaced traditional lath and plaster after World War II because it is so much lower in cost and easier and quicker to install. Over the past quarter century its use has grown at about 3 percent per year — faster than almost any other building material.

Gypsum board is made by sandwiching gypsum, a naturally occurring mineral, between a covering material, usually paper. Gypsum, deposited as the water in ancient inland seas evaporated, is a finite resource but in no danger of running out. About 65 percent of the wallboard manufactured is the conventional paper-faced variety. Also available are water-resistant wallboard, or green board, designed for moist areas such as tub and shower enclosures and as a substrate for tile. Type X wallboard has a modified core that makes it more fire resistant. It is used in fire-rated partitions and ceilings.

Wallboard is typically applied to wood studs or metal studs with nails, screws, and/or adhesives. The joints where the wallboards meet are concealed with joint tape and then finished with joint compound.

Conventional Products

The major components of conventional wallboard are gypsum and paper. The gypsum is mined in the United States or imported from Canada and Mexico. After the gypsum is extracted, usually from open pit mines, it is "calcined," or heated to remove 75 percent of the water. On the back side of most gypsum board is a gray kraft paper, usually comprised largely of recycled wastepaper, which requires fewer chemicals and less bleaching than virgin pulp. On the front side, the paper is generally a lighter cream color.

Health and Environmental Concerns

Papermill Pollution

Because papermaking is ranked fourth in total discharges of toxic pollutants by the U.S. Environmental Protection Agency, it's worth noting briefly how it is done. The mostly recycled paper used in wallboard is deinked, then bleached. Chlorines are generally used in the bleaching process to whiten the paper fibers. One by-product of this process are dioxins, including the particularly toxic 2,3,7,8-tetrachloro-dibenzo-p-dioxin (TCDD), a suspected carcinogen. Discovery of dioxins in mill wastewater and sludge has led to strong pressure on the industry to reduce dioxin emissions; at least one U.S. mill expects to be producing chlorine-free pulp on a continuous basis by 1995. Although the recycled fibers used in paper for wallboard require less bleaching (and certainly less virgin wood), the toxic metals such as cadmium used in some inks in trace amounts end up in mill wastes — one good reason why environmentalists are promoting the use of nontoxic inks.

Energy Consumption

Substantial amounts of energy are consumed during the calcining of gypsum and the drying of the wallboard. Papermaking is also quite energy intensive; producing wallboard paper from recycled stock requires about 30 percent less energy than from virgin wood.

Indoor Air Quality

Studies have shown that gypsum wallboard is a significant source of volatile organic compounds (VOCs) in the home, including formaldehyde, a strong irritant and suspected cancer causer (see "A Guide to Home Pollutants," pages 245–249). Gypsum, largely inert, is extremely low in emissions; rather, the additives used to produce water-resistant green board and fire-resistant X-type wallboard and to improve the workability of the gypsum slurry during manufacture are the likely culprits. The recycled paper on each side of the wallboard may also contain chemicals from previous uses. VOCs in the joint compound add considerably to the pollution load (see "Joint Compound," pages 168–169).

Disposal

Almost all of the wallboard waste from construction and remodeling ends up at landfills.

As it becomes more and more difficult to get approval to develop new landfill sites, some landfill operators and municipal officials are going so far as refusing to accept gypsum wastes.

Environmental Products

Concerns about indoor air quality have led to the development of wallboards that do not off-gas formaldehyde and other health-threatening pollutants. To ease the pressure on existing landfills, they're also being made with recycled material.

Suppliers

Domtar Gypsum

24 Frank Lloyd Wright Drive, Ann Arbor, MI 48105; (313) 930-4700
Gyproc gypsum board containing varying percentages of recycled scrap wallboard as well as gypsum created as a by-product of fossil-fuel burning and other industrial processes.

Eternit, Inc.

Excelsior Industrial Park, P.O. Box 679, Blandon, PA 19510; (800) 233-3155 or (610) 926-0100
Eterboard, a calcium silicate board.

Homasote Company

P.O. Box 7240, West Trenton, NJ 08628-0240; (800) 257-9491 or (609) 883-3300
N.C.F.R. Homasote fire-rated fiberboard made of 100 percent recycled wood fiber. Also DesignWall, Homasote fiberboard wrapped with fabric; Nova Cork, Homasote fiberboard covered with cork; and Burlap Panels, Homasote fiberboard faced with burlap.

Louisiana-Pacific

III SW 5th Avenue, Portland, OR 97204;
(800) 365-7672 or (503) 221-0800
FiberBond fiber-reinforced gypsum board made
from gypsum, perlite, and recycled newsprint
and telephone directories.

Meadowood Industries

33242 Red Bridge Road, Albany, OR 97321;
(503) 259-1303
Meadowood wall and ceiling board made from
ryegrass straw, an agricultural waste product.

Niagara Fiberboard, Inc.

P.O. Box 520, Lockport, NY 14095; (716)
434-8881
Monowall boards made from recycled wood
fiber for use in mobile homes.

Along with gypsum board, joint compound is one of the most ubiquitous building products. A typical home requires between twenty-five and fifty gallons of the "white mud."

When sheets of gypsum board are installed, joint compound is used to fill the gaps where they meet. The screws or nails used to fasten the wallboard to the studs are covered as well. Screws and nails are covered by joint compound directly, while the joints are typically covered first with paper or, less often, fiberglass tape.

Conventional Products

Most joint compound comes in a ready-made semi-liquid state. All you have to do is open the container and mix with a power tool or by hand until the compound is smooth, creamy, and uniform in texture. Contractors use special trowels to apply it, normally three coats with approximately twenty-four hours of drying time between each coat.

Joint compound is made primarily from calcium carbonate (limestone), or to a lesser extent gypsum, which give it body. Polyvinyl acetate or natural or synthetic starches are used as a binder and mica, clay, talc and/or perlite as filler. Ethylene glycol is added to control the drying time. Anti-bacterial and anti-fungal agents are also added.

Limestone, gypsum, mica, clay, talc, and perlite are all natural minerals in plentiful supply. Benzene, the preferred solvent used in the production of polyvinyl acetate, is a known human carcinogen. Vinyl acetate is regulated as a hazardous substance by the U.S. Environmental Protection Agency. Small amounts of these pollutants, within regulated limits, are released during the manufacture of polyvinyl acetate. Ethylene glycol has been replaced by reportedly less-hazardous chemicals in some formulations, but they are proprietary, and the manufacturers don't divulge what the substitutes are. The fungicides and bacteriacides, including formaldehyde, can also be quite toxic.

Health and Environmental Concerns

Indoor Air Quality

Inside your home, the primary concern about joint compounds is that the petrochemicals from which they are made release more than twenty-five volatile organic compounds (VOCs), according to a study conducted at Harvard University for the U.S. Environmental Protection Agency; as many as six of these are suspected human carcinogens (see "A Guide to Home Pollutants," pages 245–249). And, although previous studies showed that VOC emissions tend to decrease after an initial curing period, the Harvard study indicates that off-gassing of pollutants can *increase* over time under certain temperature and humidity conditions.

Environmental Products

A few manufacturers offer joint compound in powdered form, which emits far lower levels of VOCs than conventional ready-made products. From an environmental point of view the best alternative also includes natural binders instead of polyvinyl acetate, and no biocides are added.

Suppliers

AFM Enterprises, Inc.
1960 Chicago, Suite E7, Riverside, CA 92507; (909) 781-6860
AFM Joint Compound, a pre-mixed product, is a vinyl acetate emulsion like conventional joint compounds, but it has been specially formulated to emit very low levels of VOCs.

M-100 HiPO Joint Compound, Murco Wall Products
300 NE 21st Street, Fort Worth, TX 76106; (817) 626-1987
This low-toxicity joint compound contains no fungicides or preservatives and is made with natural binders. It comes in powdered form in twenty-five-pound bags and must be mixed with water. To each bag of M-100, approximately 1.5 gallons of water are added and mixed to a rather stiff consistency for five minutes in a mechanical mixer. Then another gallon of water is added and mixed for another five minutes. If not prepared as directed, the surface can be gritty and require more sanding than ordinary products. As with conventional joint compounds, three coats are best. Because this product doesn't contain fungicides, mildew may develop in moist rooms, although the Masters Corporation (Paul Bierman-Lytle's design and building firm) has used this product for more than eight years and never encountered this problem. Only what can be used in one day should be mixed; in dry form M-100 has a shelf life of six months. Other dry powder joint compounds are available at home centers and hardware stores. Check the label or contact the manufacturer for ingredients.

CARPETS AND UNDERLAYMENTS

Carpets can make a room feel warm and cozy and look luxurious. They also absorb noise like no other flooring. Carpeting, an $8.5 billion a year business, accounts for more than half of all floor coverings purchased in this country.

This wasn't always the case. Until early in the nineteenth century, when the industrial revolution spread to textile mills in New England, plain wood floors, grass matting, or canvas floor cloths were the most common floor coverings. Only the wealthy could afford carpets, which had to be imported. By 1830, advances in wool spinning were getting the fledgling American carpet industry off the ground. By midcentury, carpeting had become a middle-class amenity.

Conventional Carpeting

Until quite recently, carpets were made only from natural fibers such as wool and cotton. Today, although wool is the most prestigious carpet fiber, its market share has plummeted to about 2 percent. Synthetic fibers produced from petrochemicals have turned the industry upside down. About 80 percent of all carpet fibers used by U.S. manufacturers are nylon. Olefin, or polypropylene, and polyester account for most of the rest.

Nylon is produced from benzene, a derivative of crude oil. Chemical engineers developed the first nylon fiber for commercial use in 1958. Olefin is made from propane, a by-product of oil refining, combined with ethylene. It's a popular fiber for indoor-outdoor carpets. Polyester made its debut as a carpet fiber in the 1960s. It is produced when paraxylene, a benzene-like chemical, is combined with ethylene glycol. Polyester takes color well, and has a luxurious feel for a synthetic fiber. Synthetic carpet fibers are relatively inexpensive, durable, and resistant to stains and mildew.

Carpet Components

Carpeting consists of more than just the fibers. Backing is needed on tufted (as opposed to woven) carpets to prevent buckling. It also prevents fiber loss and fuzzing, adding to the life of the carpet. The backing is usually made of polypropylene laminated to the carpet using the petrochemical styrene butadiene rubber latex.

An additional layer of material is placed under a carpet to provide resiliency and insulation, deaden sound, prolong wear, and prevent slipping. This is known as carpet cushion, or underlayment. Urethane foams are the most common forms of carpet cushion.

Carpet tiles, and sometimes wall-to-wall carpeting, are glued down with a styrene butadiene rubber adhesive.

Manufacturers routinely treat carpet fibers with a variety of chemicals to make them less flammable, resist stains, and retain dyes, and to minimize static electricity and prevent mold and mildew. Other so-called finishing chemicals make the carpet fibers more flexible and resilient.

Health and Environmental Concerns

Indoor Air Quality

One of the most troubling aspects of conventional carpeting is its effect on the air quality inside your house. You don't have to be hypersensitive to chemicals to react badly to the smell of a newly carpeted room. Most commercial carpets release the chemical 4-phenylcyclohexene (4-PC), which is responsible for this new carpet odor. 4-PC, a suspected chemical irritant at extremely low levels of exposure, is a constituent of the styrene butadiene rubber latex typically used for carpet backing, adhesives, and binding agents for cushions.

A recent study of pollutants emitted by four typical kinds of synthetic carpets done by the Lawrence Berkeley Laboratory for the Consumer

Product Safety Commission detected forty volatile organic compounds (VOCs), many of them suspected toxins, some of them carcinogens.

According to Hal Levin, editor of *Indoor Air Bulletin* and a leading indoor air expert, VOC emissions from most carpet usually decline rapidly during the first day or two after installation. And they are relatively small compared to those from the many other sources found indoors — including the emissions from the adhesives used to install carpeting that is glued down, which are often one hundred to a thousand times higher than those used in the carpets themselves. In a poorly ventilated space, these pollutants could easily reach levels known to cause health problems. And although carpet emissions may generally be low, the impact on human health and comfort of any one VOC or of the total emissions may still be significant. Perhaps most worrisome of all is the fact that little is known about the health effects of most of these substances.

Carpets not only release their own pollutants but also can act as a "sink" for pollutants emitted by other sources, as well as for odors and dirt, dust, dust mites, dander, and other allergens. This is especially troubling because infants and small children frequently come face to face with carpeting and inhale three times the pollutant load of adults per body weight. These pollutants and particles can also be re-released to the air at a later time.

Water Consumption

Carpet manufacture uses an astonishing amount of water — an estimated 15 gallons per square yard of carpet. In 1990 an estimated 1.1 billion square yards of carpet were made in the U.S., which means that 16.5 billion gallons of water were consumed.

During the 1980s, water use increased exponentially with the increased popularity of Saxony carpets. These carpets are batch dyed, a process that requires three or four times more water per pound of carpet than continuous dyeing. The new stain-blocking agents used in recent years also have increased water consumption.

Water Pollution

Many carpet mills release large amounts of wastewater into nearby lakes, streams, and rivers. This wastewater contains dyes, mordants (used to bind the pigments to the fibers), and bleaching and finishing chemicals. These substances increase oxygen demand in surface waters, robbing it from aquatic life, and introduce toxins into the aquatic environment.

Air Pollution

Air pollutants released during the production of styrene butadiene rubber latex are a cause for concern, particularly for the communities near the manufacturing plants. In a list published by the EPA in 1990, nine of fourteen U.S. plants producing styrene butadiene latex were classified as "potentially high risk" because they could pose a very high public health threat to surrounding communities; indeed, the cancer risk posed by these plants was more than a thousand times higher than that considered acceptable by most policymakers.

Oil Depletion

Unlike natural carpeting made of wool, goat hair, cotton, jute, and india fiber, synthetic carpeting is derived from nonrenewable fossil fuels.

Ozone Depletion

Chlorofluorocarbons (CFCs), tied to the depletion of the protective ozone layer of the upper atmosphere, have traditionally been used as blowing agents to produce the urethane foams used as carpet cushion. The CFCs were released to the atmosphere as a waste product of this process. According to the Carpet Cushion

Council, a trade group, very few, if any, carpet foam cushion manufacturers still use CFCs. However, some have turned to methylene chloride as a substitute, a gas that also damages the upper ozone layer, though reportedly less than CFCs do.

Disposal

Compared to other floor coverings, carpeting is short lived. It usually lasts less than ten years and then ends up at the local landfill. Unlike natural materials, synthetic fibers, backings, and cushions do not biodegrade. Theoretically, nylon, olefin, and polyester, thermoplastic polymers, could be melted down and made into new fibers, but no used synthetic carpeting currently is being commercially recycled. Polyurethane foam cushions could also be recycled into new cushions, but this is not yet being done.

Environmental Products

All carpets, even those made entirely of natural materials, harbor dust, dust mites, and other allergens. What's more, even natural fibers absorb pollutants being emitted by other materials in your house, and can re-release them into the air at any time. Another drawback is the fact that carpeting of any kind has a short life span and therefore contributes to landfill problems. The best course of action is to keep carpeting to a minimum. That means avoiding wall-to-wall carpeting; area rugs consume fewer resources and can go outdoors for a good airing and shaking.

From the health and environmental perspective, the best carpet is made of time-tested natural fibers such as wool, cotton, jute, or goat hair. It is undyed and untreated and has a natural rubber latex backing or, in the case of woven cotton, no backing. This kind of carpet relies on renewable resources, doesn't offgas potentially toxic fumes, consumes less water during production, and generates less — and less toxic — wastewater. The exception is cotton, which can contain insecticides and herbicides used during farming. In addition, cotton farming can be detrimental to soils. As for underlayment, the environmentally preferable materials are india fiber and jute, or natural rubber cushion. Ozone-depleting CFCs and health-threatening styrene butadiene rubber latex are avoided altogether.

How do the natural alternatives compare with the commonly used synthetics? Natural fibers have a warmth and luxuriousness missing from their petrochemical counterparts. However, they cost more and are apt to absorb moisture and stains and mildew more easily, and thus are harder to maintain.

Other environmental choices are synthetic carpeting that is specially designed to keep indoor air pollutants to a minimum, and synthetic carpeting with fibers made of recycled material.

If you must use conventional synthetic carpeting, avoid those that are made with styrene butadiene rubber latex. Steer clear, too, of any foam cushion manufactured with ozone-depleting CFCs.

There are two basic ways to install carpet: stretched-in, tacked-down installation and glue-down installation. In the former, carpet is stretched over a separate cushion and tacked down around the perimeter of the room. Glue-down installation can be single or double glue-down. Single glue-down is used when a carpet has a secondary fabric or latex backing; in double glue-downs, a separate cushion is glued down first, then the carpet is glued on top of the cushion. Using stretched-in, tackless strip installation instead of glue-down can substantially reduce potentially toxic air emissions. If this isn't an option, at least avoid double glue-down, and use one of the new low-VOC adhesives now available (see "Glues and Adhesives," pages 199–201).

Suppliers

NATURAL CARPETING

Bremworth Carpets
1940 Olivera Road, Suite C, Concord, CA
94520; (800) 227-3408
Woven wool carpeting with jute backing.

Carousel Carpets
1 Carousel Lane, Ukiah, CA 95482; (707)
485-0333
Natural fiber carpets with natural latex backings.

Corniche Carpet Mills
201 Covington Street, Oakland, CA 94605;
(510) 568-8610
Untreated and undyed wool rugs in seven natural colors ranging from beiges to grays. Natural rubber latex backings.

Dellinger
1943 North Broad Street, Rome, GA 30162-0273; (706) 291-7402
All cotton and wool carpeting without chemical treatment.

Desso Carpet
P.O. Box 1351, Wayne, PA 19087; (800) 368-1515
Woven wool carpet with jute backing.

Helios Carpet
P.O. Box 1928, Calhoun, GA 30703; (800)
843-5138
Natural, untreated carpets with jute backing.

Hendrickson Naturlich
8031 Mill Station Road, Sebastopol, CA
95472; (707) 829-3959
New Zealand wool carpets.

Jack Lenor Larsen
41 East 11th Street, New York, NY 10003;
(212) 674-3993
Natural fiber carpeting.

Gordon T. Sands
40 Torbay Road, Markham, Ontario, Canada
L3R 1G6; (416) 475-6380
Woven wool carpeting.

Spectra, Inc.
1943 North Broad Street, P.O. Box 273,
Rome, GA 30162-0273; (706) 291-7402
Untreated wool and cotton carpeting in a variety of textures, including shag and sisal. Natural hues and pastel color dyes are available. Natural rubber latex backings are also available.

LOW-POLLUTANT SYNTHETIC CARPET

Collins & Aikman, Floor Coverings Division
P.O. Box 1447, 311 Smith Industrial Boulevard, Dalton, GA 30722-1447; (800) 248-2878
Powerbond RS carpeting line does not require wet adhesives, which release potentially health-threatening pollutants. A plastic membrane is stripped away during installation to expose a dry tackifier that is incorporated into the carpet backing. Studies have shown the carpeting emits no detectable 4-PC. Comes in a variety of colors. C & A carpets are 100 percent vinyl, which means there is no need for synthetic backing. Also, the carpet is vapor-proof, which means it can seal out formaldehyde released from subfloors. This also eliminates the need to remove potentially health-threatening asbestos and vinyl tile or sheet floors. However, because it is made of vinyl, this carpet fails to address environmental hazards associated with vinyl production and fossil fuel depletion.

Sutherland Carpet Mills
707 Sierra Bonita Drive, Placentia, CA 92670;
(714) 447-0792
Low-VOC carpeting.

RECYCLED CARPET
Image Carpets, Inc.
Box 5555, Armuchee, GA 30105; (800) 722-2504 or (404) 235-8444
All or most of the fiber of the carpeting in its Enviro-Tech line comes from recycled PET plastic soda bottles. (Thirty-five two-liter bottles make one square yard of carpet). The fibers are colored with synthetic dyes and treated with biocides. The backing is styrene butadiene rubber latex. From the standpoint of potential indoor air pollution, this isn't the best choice, but its recycled content is a plus.

UNDERLAYMENT
Chris Craft Industrial Products, Inc.
P.O. Box 70, Waterford, NY 12188; (800) 765-4723 or (518) 237-5850
One hundred percent jute fiber carpet cushion and several different cushions made of recycled carpet and textile fibers. These are treated chemically for fire retardance and moisture and mildew resistance.

Dodge-Regupol
P.O. Box 989, Lancaster, PA 17608-0989;
(800) 322-1923 or (717) 295-3400
Carpet underlayment made of 100 percent recycled tire rubber.

Dura Undercushion
8525 Delmeade Road, Montreal, Quebec,
Canada H4T 1M1; (514) 737-6561
Padding made from recycled rubber.

Homasote Company
P.O. Box 7240, West Trenton, NJ 08628-0240;
(800) 257-9491 or (609) 883-3300
440 CarpetBoard and Comfort Base carpet underlayments are made from 100 percent recycled newsprint. They are treated with a fungicide.

R. B. Rubber Products
Contact John Whitney, 904 East 10th Avenue, McMinnville, OR 97128; (800) 525-5530 or (503) 472-4691
Rubber matting made from 100 percent recycled tire rubber.

Gordon T. Sands
40 Torbay Road, Markham, Ontario, Canada L3R 1G6; (416) 475-6380
Felt carpet padding.

Sutherland Carpet Mills
707 Sierra Bonita Drive, Placentia, CA 92670;
(714) 447-0792
Low-VOC carpet padding.

"Resilient flooring" is the term used to describe products like linoleum, rubber, and vinyl tile and sheet flooring. They're called resilient because they spring back into shape and therefore are very resistant to dents — and quite comfortable to walk or stand on. They're also very durable and easy to maintain.

In the 1860s the Scottish inventor Frederick Walton developed the first resilient flooring, linoleum. Walton obtained an American patent and in 1875 began producing linoleum in New York as the American Linoleum Manufacturing Company. Ella Rodman Church, author of *How to Furnish a Home* (1882), sung the praises of the new linoleum as warmer, more durable, and more reasonably priced than the painted floor cloths that were the standard floor covering of the day. The 1897 Sears, Roebuck catalog described it as very similar to floor cloth "except that there is ground cork in its composition, which makes it much heavier, more durable; also, very much softer to walk on." By the end of the century, linoleum was the popular choice.

Today, when most people think of linoleum they think of vinyl. However, real linoleum is made entirely of natural ingredients. It has been replaced by vinyl floorings and asphalt tiles only since the end of World War II.

Conventional Flooring

Virtually all resilient flooring today is made of vinyl, whether sheets or tiles. Polyvinyl chloride (PVC) is the main component of vinyl flooring, derived primarily from petroleum and chlorine feedstocks.

Plasticizers, or softeners, are required for the manufacture of flooring from PVC; according to the American Institute of Architects' *Environmental Resource Guide*, vinyl flooring may contain as much as 20 percent plasticizer.

One plasticizer is 2,2,4-trimetyl-1,3-pentanediol-diisobutyrate, also known under the trade name TXIB. Sheet vinyl contains more plasticizer than vinyl tile and is therefore more flexible.

Vinyl flooring is extremely resistant to dents and grease stains. It's comfortable underfoot and very easy to maintain — indeed, most of the vinyl flooring sold today has a urethane coating and doesn't need to be waxed. Nearly 75 percent of new homes have sheet vinyl flooring in the kitchen. Vinyl is also commonly used in bathrooms, laundry rooms, and recreational rooms. It lasts fifteen years or more.

The lowest-priced synthetic flooring available today is asphalt tile. Asphalt tiles, developed in the 1920s, are made of 50 percent limestone or other mineral fillers and 25 percent asbestos, all held together with asphalt, which comprises the remaining 25 percent. (Since 1986, asbestos has been banned in U.S. tiles. However, asphalt tiles are no longer made in this country but rather are imported.) Asphalt tiles are brittle, not very resistant to dents, and difficult to maintain, but they're fairly comfortable and quiet underfoot.

Health and Environmental Concerns

Indoor Air Quality

Some of the petrochemical constituents of vinyl flooring contain volatile organic compounds, or VOCs (see "A Guide to Home Pollutants," pages 245–249), and can pollute the air inside your house. One study of emissions from vinyl flooring identified twenty-seven compounds, five of them suspected cancer causers. A Finnish study of seven vinyl flooring samples found high VOC emission rates in all of them.

The plasticizer in vinyl flooring is the main source of emissions. According to a Swedish study, the plasticizer TXIB was a frequent cause of health complaints.

The offgassing of these pollutants decreases substantially twenty four hours after installation — if the space is well ventilated. However, low emissions of TXIB and other VOCs from vinyl flooring can be detected years after installation, adding to the pollutant load inside your home.

Toxicity

In 1989, 380 million pounds of polyvinyl chloride were used in the manufacture of vinyl flooring. From the health and environmental point of view, PVC manufacture is one of the most hazardous businesses around. PVC is made from vinyl chloride, a known carcinogen. Because PVC is considered a "priority pollutant" by the federal Clean Water Act, discharges from vinyl chloride and polyvinyl chloride manufacturing plants are subject to special wastewater discharge permits. By the same token, air emissions from these operations are strictly regulated, and some solid wastes from PVC manufacturing are classified as hazardous and must meet special regulations for handling and disposal. PVC is being banned in parts of Europe.

A variety of other chemicals considered hazardous by the federal government may also be found in vinyl flooring, including formaldehyde, a probable human carcinogen.

Vinyl flooring used to contain asbestos, but since the mid-1980s, asbestos has been banned. If you're remodeling or renovating a house that has a vinyl floor installed before then, keep this in mind. The best thing to do is to install an airtight new flooring over the old one. If you must remove it, use a professional trained in asbestos removal.

Asphalt tiles also emit volatile organic chemicals. Imported tiles may also contain asbestos. In any event, any asphalt tiles installed before 1986 did contain asbestos and must be handled cautiously, as the vinyl floors described above.

Oil Depletion

Vinyl flooring, as well as asphalt tiles, are made from dwindling supplies of fossil fuels.

Disposal

Little used vinyl flooring is recycled in this country; once it's ripped out it generally goes to a landfill (technically, pre-1986 vinyl flooring, which contains asbestos, must be handled as a hazardous waste and disposed of in authorized landfills). It is not biodegradable. Technically, vinyl flooring is recyclable; in fact, in Europe recycling systems are being developed. Here in the United States, however, no collection and recovery systems are in place.

Environmental Products

Linoleum and cork have been used for decades, and they're still the healthiest and most handsome choices for resilient flooring. They are made of natural materials that are constantly renewed by nature. They produce far less indoor air pollution than their synthetic counterparts, and are less toxic to manufacture. What's more, they are both biodegradable and recyclable. They also rival the synthetics in durability and ease of maintenance.

Linoleum

Linoleum is made primarily from 100 percent natural powdered cork, linseed oil (made from flax seed), wood resin, and wood flour pressed onto a backing of burlap or jute. The linseed oil is slowly oxidized and mixed with the wood resin in large tanks, Oxygen is forced through the tanks for twenty-four to thirty hours until a sticky, jelly-like material called "cement" forms. The jelly-like slabs of cement are mixed with wood dust, cork, and pigments. Then granules of various colors are blended together and passed through rollers to produce sheets that are pressed onto burlap or jute backing. Finally, the sheets are hung in huge, heated halls for up to three weeks to cure.

Linoleum has all the pluses of vinyl flooring and none of the minuses. It is extremely durable. In fact, ten-year-old linoleum is actually tougher than newly laid linoleum because the linseed oil continues to oxidize, forming new chemical links. The wear on a piece of linoleum laid in 1938 suggests that the flooring could last for hundreds of years!

Linoleum floors resist indentation, abrasions, and cracking. They are comfortable underfoot in more ways than one: unaffected by floor temperatures, they're always pleasantly warm. They are greaseproof, waterproof, and termite proof. They don't burn easily and will not emit dangerous gases when exposed to heat. They do not contribute substantially to indoor pollution; in fact, they actually help prevent it because the continual oxidation of the linseed oil inhibits the growth of bacteria.

Linoleum hasn't been produced in the United States since the mid-1970s but is still available from three European companies, including the original Scottish manufacturer. It is available as either sheets or tiles in many patterns, colors, and textures. Rolls are six feet and wider. Tiles come in 12-inch squares and are 1/8-inch thick.

Linoleum does require more energy to manufacture than vinyl flooring (49,934 Btu per pound of product, as opposed to 34,000 Btu/lb for vinyl, according to a study by Franklin Associates, Ltd.). It is also more expensive.

Cork

The use of cork dates back more than three thousand years. The Romans used cork to seal wine and oil amphoras.

Cork flooring is produced from the outer bark of the cork oak (*Quercus suber*) of southern Europe, primarily Portugal and Spain. In a process called "unmasking," the bark is stripped off the tree. The cork oak regenerates its bark after stripping, and all the bark is used without waste. The interval between strippings is regulated by law; in Sardinia, for example, it is ten years.

The bark of the cork oak has a fourteen-sided cellular structure; one cubic inch contains over 200 million cells. This distinctive structure is what gives cork its famous lightness, compressibility, and resilience. Cork is made by heating the bark, causing the granules to adhere to each other without additional adhesives.

Cork flooring is generally available only as tiles, although some cork sheeting, or carpet, is made from cork granules and linseed oil with a jute canvas backing. The tiles come in 12-inch squares, and in two thicknesses, 3/16 inch and 5/16 inch.

Unfinished tiles are preferable to the tiles available at commercial outlets, which are coated with a layer of vinyl or urethane. The unfinished tiles should be sealed with a natural or low-VOC synthetic sealer. Unfinished cork ranges in color from pine to walnut, but also may be tinted darker tones using a tinting oil (see "Sealers and Finishes," pages 189–197).

Cork tiles are moisture-, rot-, and mold-resistant, don't conduct heat, and have an amazing ability to absorb vibration and deaden sound. They are not as durable or easy to maintain as linoleum, requiring periodic waxing or oiling. Properly maintained, however, cork can last thirty years. Cork tile requires only occasional cleaning with a mild detergent and damp cloth. A wax cleaner is recommended (see "Home Maintenance Products," pages 238–244).

Recycled Rubber

Rubber accounts for a very small but growing percentage of the flooring market, thanks to concerns about the millions of discarded tires piling up across the country. Rubber floor tiles, like automobile tires, are usually made from synthetic rubber. They are very resilient, dent and stain resistant, comfortable, and quiet. Long lasting, they are suitable for patios, decks, walkways, and exercise areas. Don't use them indoors, however, because the recycled rubber may be a source of air pollutants.

Healthy Adhesives
Both linoleum and cork tiles must be applied to the subfloor with an adhesive. Conventional adhesives emit irritating and potentially toxic fumes. See "Glues and Adhesives," pages 199–201, for healthier alternatives.

Suppliers

CORK TILES
Bangor Cork
William and D Street, P.O. Box 125, PenArgyl, PA 18072; (215) 863-9041

Corticeira Amorim
P.O. Box 1, Mozelos 4539 Lourosa Codex, Portugal; 2-764-7509

Dodge-Regupol, Inc.
P.O. Box 989, Lancaster, PA 17608-0989; (800) 322-1923 or (717) 295-3400

EX
400 East 56th Street, New York, NY 10022; (212) 758-2593

NATURAL LINOLEUM SHEET AND TILES
Forbo Industries, Inc.
Humboldt Industrial Park, Maplewood Drive, P.O. Box 667, Hazleton, PA 18201; (800) 233-0475 or (717) 459-0771

Gerbert Limited
P.O. Box 4944, Lancaster, PA 17604; (800) 828-9461 or (717) 299-5035

Hendricksen Naturlich
8031 Mill Station Road, Sebastopol, CA 95472; (707) 829-3959

Nairn Floors International
560 Weber Street North, Waterloo, Ontario, Canada N2L 5C6; (519) 884-2602

Non-Toxic Environments
6135 NW Mountain View Drive, Corvallis, OR 97330; (503) 745-7838

RECYCLED RUBBER FLOORING
Carlisle Tire & Rubber Company
Box 99, Carlisle, PA 17013; (800) 233-7165 or (717) 249-1000

Dodge-Regupol Incorporated
P.O. Box 989, Lancaster, PA 17608-0989; (800) 322-1923 or (717) 295-3400

Cotton isn't the only plant fiber appropriate for carpeting. Other plants with durable fibers have long been used to furnish houses around the globe. Plant-fiber carpets, known in those days as matting, were very popular in early American homes both as a year-round floor covering and as a cool alternative to heavy wool carpets in summer. They remained the least expensive floor covering until the end of the nineteenth century.

Carpets made from these natural fibers are a rustic alternative to wool and cotton carpets. The various plant fibers and different weaves provide a range of interesting textures for your rooms. These rugs also have an air-conditioning effect, absorbing surplus moisture, storing it, and then releasing it into the air when it is dry.

Most plant-fiber carpets are made from a handful of plants:

- **Coir.** The husk of the fruit of the coconut palm, *Cocos nucifera,* widely cultivated in the tropics, yields these tough, coarse fibers. The fibers are so tough, in fact, that they must be immersed in salt water for months before they become pliable enough to spin into yarn. Coir is strong and hard-wearing, but somewhat scratchy underfoot. Coir rugs are most suitable for porches, sunrooms, and other rustic spaces.

- **Jute**. Jute is derived from tropical Asian plants belonging to the genus *Corchorus*. The strong, glossy plant fibers are obtained from the plants' inner bark. Jute is widely cultivated as an annual crop, particularly in India and Bangladesh. It is softer and silkier than many other plant fibers.

- **Seagrass.** Carpets known as seagrass are usually a mix of sisal and tropical reeds native to wetland environments.

- **Sisal.** Sisal comes from a Mexican agave species, *Agave sisalana,* known for the strength of its long leaf fibers. The fibers are extracted from the leaves once they are at least three feet long. Sisal is not as coarse as coir. Most people consider it the best looking of the natural plant fibers.

These natural plant fibers are available as wall-to-wall carpeting, area rugs, carpet tiles, and runners.

Conventional Plant-Fiber Carpets

Most commercially available carpets made from plant fibers have been treated with insecticides, fungicides, and mildewcides. They're also routinely treated with fire retardants and stain repellents.

Many of the carpets are a combination of plant fibers and synthetic fibers. For example, sisal boucle is sisal combined with wool-acrylic yarn. In addition, they are sometimes combined with synthetic backings (see "Carpets and Underlayments," pages 170–174).

Health and Environmental Concerns

Indoor Air Quality

The petrochemical insecticides, mildewcides, fungicides, stain repellents, and fire retardants used on most plant-fiber carpets release pollutants called volatile organic compounds (VOCs) that can make your indoor air unhealthy (see "A Guide to Home Pollutants," pages 245–249). Like all carpets, plant-fiber rugs become "sinks" for odors, VOCs, dirt, dust, dander, and other allergens that they can re-emit into the air in your rooms at any time. Consequently, small area rugs are best because they provide less area for allergens and pollutants to be absorbed.

If your plant-fiber carpeting is being glued down you have something else to think about. The adhesives used are some of the worst sources of air pollutants in the typical house.

Low-VOC adhesives are available (see "Glues and Adhesives," pages 199–201).

Water Pollution

Wastewater from the fiber bleaching and dyeing process contains pigments, mordants, bleaching agents, and other substances that rob oxygen from waterways and introduce toxins, harming aquatic life.

Disposal

Carpets that are a blend of natural plant fiber and synthetic fibers and backings are difficult to separate for recycling and are more likely to end up in already overburdened landfills.

Environmental Products

The healthiest carpet, both for you and the environment, is 100 percent untreated — no toxic insecticides and fungicides, fire retardants, or stain repellents. If it is backed, the material used for backing is natural rubber latex. Although untreated carpets are healthier, they're also harder to keep clean than treated carpet, and more flammable. Therefore, when you shop for an area rug, you should weigh such factors as the amount of traffic and abuse to which it will be subjected, possible sources of moisture in the space, and the amount of ventilation the room will receive and choose the one best suited to the particular situation.

Suppliers

Design Materials, Inc.
241 South 55th Street, Kansas City, KS 66106; (800) 654-6451 or (913) 342-9796
Sisal and coir carpets.

Environmental Construction Outfitters
44 Crosby Street, New York, NY 10012; (800) 238-5008 or (212) 334-9659
Distributors of untreated cocoa mat rugs from two German companies, Dekowe and Oschwald. Cocoa mat is made from coconut shell fibers.

Hendricksen Naturlich
8031 Mill Station Road, Sebastopol, CA 95472; (707) 829-3959
Sisal, coir, and seagrass carpets. Wool/sisal blends are also available.

Merida Meridian, Inc.
6603 Joy Road, East Syracuse, NY 13057; (800) 345-2200 or (315) 433-1111
Coir, sisal/coir blend, and seagrass carpets in hand-painted and jacquard designs. The rugs are backed with natural rubber latex.

Prestige Mills
83 Harbor Road, Port Washington, NY 11050; (516) 767-1110
Sisal carpeting with latex backing.

Alison T. Seymour
5423 W. Marginal Way SW, Seattle, WA 98106; (206) 935-5471
Sisal carpeting with latex backing.

PAINTS

The urge to paint goes back at least fifteen thousand years, when Stone Age artists decorated the walls of their caves. For centuries, only natural ingredients were used to make paints. In the fourteenth century, natural pigments were mixed with egg yolk to produce brilliant and luminous egg tempera paints. One hundred years later, most paints were made with linseed oil. The nineteenth century saw the advent of synthetic dyes and many new colors, including cadmium and chrome yellow, mauve and ultramarine. Since World War II petroleum-based synthetics have largely replaced the natural oils and resins in traditional paints.

Paint is used on almost every surface of the modern home — walls, windows, doors, ceilings, cabinets, furniture, and sometimes even floors. In 1989, 493 million gallons were used in American houses.

During the past couple of decades, manufacturers have concentrated on making painting easy for the do-it-yourselfer: today there are paints that provide one-coat coverage, dripless ceiling paints, and water-based paints for easy cleanup. Now health and environmental concerns are once again turning the paint industry upside down.

There are two basic types of paints: water-based, or latex, and oil-based. Whether oil- or water-based, paints have four basic components:

A **solvent** keeps the paint liquid and evaporates as it dries.

Binders are the resins and oils that harden, forming the durable coating. The type of resin in the paint depends on the solvent. Oil-based paints are usually alkyd resins, while water-based can be one of a handful of resins, such as acrylic latex.

Pigments include both the paint color and an opacifying substance that helps the paint to cover the surface.

Additives enhance the paint's performance. Examples are driers, thickeners, anti-skinning agents, preservatives, fungicides, and mildewcides. Although they usually comprise less than two percent of the weight of paint, additives can be quite toxic.

Paints are a lot safer than they used to be. More paints are water-based than ever before. They no longer contain lead or mercury. But there are still problems.

Paints will last longer on smooth, edge-grained lumber, while oils and stains are best applied on rough-sawn and flat-grained lumber.

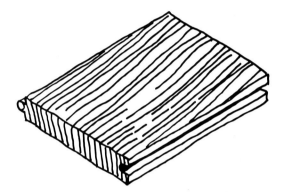

Edge-grained wood

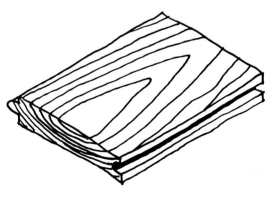

Flat-grained wood

Water-based, or latex, paint is generally used for walls and ceilings. There are several different types, depending on the binders used. The three most common are styrene butadiene, polyvinyl acetate, and acrylic resins.

Health and Environmental Concerns

Indoor Air Quality
Latex paint uses water as the principal solvent, greatly reducing the hazards associated with use. In addition, hazardous thinners are not needed for cleanup. However, even water-based paints contain some troublesome solvents, as well as a wide variety of vehicles, dispersants, stabilizers, surfactants, and preservatives that are volatile to some degree, meaning that they become breathable gases — and therefore indoor air pollutants — at room temperature. Some of these irritate the respiratory tract; some, such as acrylonitrile, are suspected carcinogens; and others, such as ethylene glycol and ethylene glycol ethylethers, are suspected hazards to fetuses. One study of thirty-one building materials listed a latex paint as the third largest contributor of volatile organic compound (VOC) emissions (see "A Guide to Home Pollutants," pages 245–249). What's more, slower drying times and therefore emissions of VOCs over longer periods of time make water-based paints less attractive than they might first appear to be. Some research indicates that water-based paints may emit glycol ethers and other pollutants more than six months after application.

Toxic Manufacturing Wastes
Styrene butadiene and other synthetic resins used to make paints are extremely toxic to manufacture. Paint manufacturers are regulated under numerous federal laws, including the National Pollution Discharge and Elimination System, which controls pollutant discharges to waterways, and the Clean Air Act. A study of several paint manufacturers in California demonstrated that manufacturers could do a much better job of reducing the quantity and toxicity of manufacturing wastes.

Oil Depletion
Even water-based paints contain materials that are derived from petroleum products.

Environmental Products

In recent years, as concerns about air pollution have risen, alternative paints have been developed to reduce harmful emissions. These new paints differ from conventional paints in various ways, with varying degrees of environmental improvements.

Natural Paints
Natural paints contain no toxic preservatives, mildewcides, or fungicides. The pigments are derived from clay and minerals and do not contain cadmium and other toxins used in conventional paints. Natural paints do not use nonrenewable petrochemical solvents or resins, or else use them in very small amounts. Instead, the solvents are derived from lemon or citrus oils.

These natural products have slightly different and more subtle hues than those made from synthetic sources. They have a strong, sweet citrus scent. They can be applied to plaster, concrete, and wallpaper and over other paints.

Two brands of natural paints are currently available in this country; and they're both made in Germany, which in 1974 passed one of the first laws that reduced allowable emissions of hazardous solvents in paints. Only one company, Auro, produces paints that are completely "natural." Because it received some complaints about the strong smell of d,l-limonene, the aromatic citrus-based solvent used in its natural paints, the other company, Eco Design (formerly Livos PlantChemistry), augments it with a small quantity of petroleum-based solvent. The U.S. Food and Drug Administration has classified d,l-limonene as an irritant and potential health hazard, although the agency has yet to qualify the significance of this health risk; consequently, it is still advisable to use these paints with caution and make sure there is plenty of ventilation where you work. Reports from Sweden, Denmark, and Germany, where natural paints have been used for years, conclude that they are quite low in VOC emissions.

Natural paints differ from their conventional counterparts not only in content but also performance. If you insist on a quick and easy product, they're not for you. These paints are, however, ideal for those who want the highest-quality finish, both aesthetically and environmentally, and are willing to take the time to do the job right. Some require more expertise to prepare and apply. They may require more coats and longer drying times than conventional paints. Some natural paints are only available in white and off-white; if you want colors you have to mix them yourself or go to a paint store with mixing equipment. Some can't be rolled, brushed, or sprayed as easily as conventional paints. Some are more sensitive to temperature and humidity. Because they lack preservatives, they have limited shelf lives.

The natural paints are sold in European metric containers, but the quart and gallon equivalents and area coverage are listed on the label. Because their natural ingredients originate from many countries around the world and must be imported, they are expensive. But they are worth every penny.

Milk Paint

Another natural paint is available for special uses. Made by the Old-Fashioned Milk Paint Company, which has re-created traditional American homemade milk paint, it is free of petrochemical solvents. It is made from milk and milk products and various minerals such as clay, lime, and colored oxides. It is mixed and hand-packed in powder form into brown paper bags complete with mixing directions; you simply add water to milk paint and stir.

Milk paints are particularly well suited to restoration or refinishing of antique furniture and historic buildings. They can be used on walls, ceilings, floors, and woodwork, and can be applied with brush, roller or spray. Thickness or transparency can be controlled with the quantity of water used in mixing. Milk paints are available in twelve colors.

Whitewash Paints

The two German natural paint companies mentioned above offer whitewash paints, which are quite beautiful on plaster, concrete, and adobe. They are also reasonably priced.

Low-Biocide Paints

Low-biocide paints are conventional paints with an important difference: 90 to 95 percent of the preservatives and fungicides have been omitted. This reduces pollutant levels substantially; although these paints are made with petroleum-based chemicals, they meet California's strict limitations on VOC emissions. Unlike most paints, they have very little odor, even when they're first applied — so little, in fact, that the chemically sensitive generally can tolerate them.

Fungicides are added to most paint to control mildew, a common problem in humid areas. For these areas, the oil-based enamels discussed in the next section are a better option. If you use low-biocide paints and you live in a humid climate, make sure your rooms have plenty of daylight and are well ventilated, either naturally or mechanically.

Low-biocide paints handle a lot like conventional paints and are therefore easy to use. They can be rolled, brushed, and sprayed more easily than the natural paints. A range of pre-mixed colors are available. They are sold in the familiar gallon containers. Prices are more competitive with conventional paints. On the other hand, because they contain little or no preservative, they don't last more than about six months; don't expect to be able to store your leftover paint for use next year. This also means that the paints must be made to order. Expect a four- to eight-week delivery time.

Low-VOC Paints

These paints are specially formulated to be low in polluting emissions. Low-VOC paints may not be totally free of hazardous constituents, but they have been formulated to omit certain substances and substitute others that are less harmful. For example, petroleum-based solvents such as ethylene glycol, which used to make up about 10 percent of the product, have been reduced to a few percent or even to zero. Low-VOC paints handle a lot like conventional paints.

Suppliers

NATURAL PAINTS

Auro, Sinan Co.
P.O. Box 857, Davis, CA 95617-0857; (916) 753-3104

Eco Design
1365 Rufina Circle, Santa Fe, NM 87501; (800) 621-2591 or (505) 438-3448

MILK PAINT

Old-Fashioned Milk Paint Company
Box 222, Groton, MA 01450; (508) 448-6336

WHITEWASH PAINT

Auro Natural Chalk Water Paint
Box 857, Davis, CA 95617-0857; (916) 753-3104

Eco Design Albion White Wash Paint
1365 Rufina Circle, Santa Fe, NM 87501; (800) 621-2591 or (505) 438-3448

LOW-BIOCIDE PAINT

Miller Paint Co.
317 S.E. Grand Avenue, Portland, OR 97214; (503) 233-4491

Murco Wall Products
300 N.E. 21st Street, Fort Worth, TX 76106; (817) 626-1987

LOW-VOC PAINT

AFM Enterprises
1140 Stacy Court, Riverside, CA 92507; (714) 781-6860

Bonakemi
14805 East Moncrieff Place, Aurora, CO 80011-1411; (800) 872-5515

Ecos Paint
P.O. Box 375, Saint Johnsbury, VT 05819; (802) 748-9144

Glidden Spred 2000
925 Euclid Avenue, Cleveland, OH 44115; (800) 221-4100

Kurfees Paints
201 East Market Street, Louisville, KY 40202; (502) 584-0151

Pace Chem
P.O. Box 1946, Santa Ynez, CA 93460; (800) 350-2912

Pristine Clean Air Formula, Benjamin Moore Paints
51 Chestnut Ridge Road, Montvale, NJ 07645; (201) 573-9600

Weather Bos
1774 Rainier Avenue South, Suite 130, Seattle, WA 98144; (206) 329-3663

ENAMEL PAINTS

Some painting jobs require a tough moisture- or weather-resistant coating — trim, window sash, doors, or areas with high moisture such as bathrooms, greenhouses, and exercise rooms. This is especially true of exterior surfaces exposed to the weather. For these jobs, enamel paints fit the bill. Enamel paint usually means a hard, high-gloss paint. Enamel used to be only oil-based but now is available in water-based latex formulations as well.

Conventional Products

For a discussion of water-based paints, see the preceding section. In oil-based paint the main hazard comes from the solvents. Whereas water-based paints contain from 5 to 15 percent petrochemical solvents by weight, oil-based versions contain about 50 percent. A large number of different solvents can be present in a single formulation, many of them considered health hazards by the federal government: among the solvents used are aliphatic hydrocarbons, glycols, ketones such as acetone, methyl ethyl ketones and methyl isobutyl ketone, and aromatic and naphthenic hydrocarbons, including toluene, o-xylene, naphtha, and ethyl acetate — all hazardous chemicals. Polyester resins (alkyds), polyurethanes, and silicones, which all derive from petroleum, are the binders, in addition to large amounts of linseed and other natural oils. There are also toxic additives such as styrene and vinyl toluene for toughness, acrylic monomers for gloss, adhesion, and durability, and various formaldehyde compounds for hardness, gloss, and stain resistance. The same pigments used in water-based paints, some of which are highly toxic, are used in oil-based paints as well. However, oil-based enamels generally do not contain preservatives and fungicides because they are inhospitable to bacteria and molds.

Health and Environmental Concerns

Take the health and environmental effects of water-based paints discussed in the previous section, multiply them many times, and you have some idea of the hazards of oil-based enamels.

Indoor Air Quality
Oil-based enamels are an even greater health threat in your home than water-based paints because they emit far more volatile organic compounds.

Air Pollution
The VOCs don't just pollute the air inside your home. They are also photochemically reactive — that is, they react with the ultraviolet rays in sunlight to produce ozone, a major component of smog. These airborne emissions are coming under stricter regulation by the 1990 amendments to the federal Clean Air Act, forcing paint manufacturers to substantially reduce VOC emissions from their products. A few states such as California are coming up with their own, stricter, emission limits.

Oil Depletion
Even more nonrenewable fossil fuel goes into the production of oil-based enamels than is used in the making of water-based paints.

Energy Consumption
More energy is consumed in the manufacture of oil-based enamels than in that of water-based paints as well — 41,642 Btus per pound compared to 33,034 Btus/lb., according to a Finnish study.

Disposal
Conventional oil-based paints and their containers are classified as hazardous waste in some communities.

Environmental Enamels

The environmental first choice are German enamel paints made with mostly natural ingredients. The same natural solvent, d,l-limonene, used in the natural paints discussed in the previous section, is also in these products. Most people find the resulting fragrance very pleasant. However, the U.S. Food and Drug Administration lists this citrus solvent as an irritant, so be sure there's adequate ventilation while you work. A citrus thinner is used with the German paints for cleanup. Specially formulated low-VOC synthetic enamels are another alternative.

The German-made products provide by far the most beautiful finish. Use them on wood trim, window sash, doors, furniture, cabinets, shelving, pantries, and any wood surfaces exposed to the weather. The low-VOC paints are good choices for larger areas subject to humid conditions.

Suppliers

NATURAL PAINTS
Auro, Sinan Co.
Box 857, Davis, CA 95617-0857; (916) 753-3104

Eco Design
1365 Rufina Circle, Santa Fe, NM 87501; (800) 621-2591 or (505) 438-3448

LOW-VOC PAINTS
AFM Enterprises
1140 Stacy Court, Riverside, CA 92507; (714) 781-6860

Bonakemi
14805 East Moncrieff Place, Aurora, CO 80011-1411; (800) 872-5515

Ecos Paint
P.O. Box 375, Saint Johnsbury, VT 05819;
(802) 748-9144

Glidden Spred 2000
925 Euclid Avenue, Cleveland, OH 44115;
(800) 221-4100

Kurfees Paints
201 East Market Street, Louisville, KY 40202;
(502) 584-0151

Miller Paint Co.
317 S.E. Grand Avenue, Portland, OR 97214;
(503) 233-4491

Murco Wall Products
300 N.E. 21st Street, Fort Worth, TX 76106;
(817) 626-1987

Pace Chem
P.O. Box 1946, Santa Ynez, CA 93460; (800)
350-2912

Pristine Clean Air Formula, Benjamin Moore Paints
51 Chestnut Ridge Road, Montvale, NY
07645; (201) 573-9600

Weather Bos
1774 Rainier Avenue South, Suite 130, Seattle,
WA 98144; (206) 329-3663

In the American home, wallpaper has gone in and out of fashion. The American colonists used stencilling to embellish their walls; until the mid-nineteenth century, only the wealthy could afford wallpapers. During the High Victorian period, richly colored wallpapers and the "tripartite" wall came into vogue: different papers were used for the wainscoting at the bottom of the wall, the frieze or cornice at the top, and the field in between. Elaborate patterns were all the rage: peacocks strutted across cornices, and stylized plant motifs were popular for fields. For years, stark white walls were chic in modern interiors, and wallpaper was relegated mostly to the bathroom and the kitchen. In the neo-Victorian interiors of the 1980s, however, wallpapers and wallpaper borders made a strong comeback.

Health and Environmental Concerns

Wallcoverings and the adhesives used to attach them can be major sources of indoor air pollution. Popular vinyl "wallpapers," really plastic films, emit toxic vapors, including vinyl chloride, a carcinogen. And, like other plastics, they are made of nonrenewable fossil fuels. Even wallcoverings made of fabric or paper are dyed with chemical inks and treated with mildewcides, fire retardants, and stainguards, all of them volatile organic compounds, or VOCs (see "A Guide to Home Pollutants," pages 245–249), that can make the air inside your home unhealthy to breathe. Metallic foil wallcoverings may be less polluting than plastic or treated paper and fabric but do not represent a wise use of metals such as aluminum, which must be mined from rainforest areas and requires a great deal of energy to produce.

Even worse than the wallcoverings themselves are the adhesives used to glue them to the walls. Unlike wheat paste mixed with water, the traditional wallpaper glue, today's adhesives are generally petroleum-based synthetic formulations and potentially large sources of VOCs. Wallpapers and adhesives are particularly troublesome because they cover such a large area in your rooms, and therefore can be large sources of toxic emissions.

Environmental Products

Wallcoverings not only can pollute your indoor air but are also an unnecessary waste of resources, whether paper, plastic, or metal. Whenever you're tempted to use them, think about whether they're really necessary. Instead, why not use natural or low-VOC paints to stencil a border, or create a faux finish on your walls?

If you do decide to paper your walls, look for untreated natural wallcoverings made of paper, linen, or other plant fibers. Because these are not mass produced, they are expensive. Cork wallpaper, thin sheets of natural cork from Sardinia, is another possibility. It is beautifully textured, providing a leather-like finish, but very fragile and consequently most suitable for a library or study, not active areas or those likely to be frequented by children.

Much more widely available, and affordable, are wallcoverings made of gypsum-coated fabric or pure glass yarns. These have several environmental advantages: they are made from gypsum, a naturally occurring mineral, or in the case of glass, quartz sand, soda lime, and dolomite — all natural materials in virtually unlimited supply. These minerals are inert and emit no toxic fumes. They're also extremely durable, don't burn or rot, and provide no breeding ground for microorganisms. Like grass cloth, they come in a variety of textures, and they can be applied to almost any surface, including wallboard, cinderblock, and plaster.

Be sure to apply any wallcovering only with a natural or low-VOC adhesive. Natural wallpaper glues are generally made of water and plant starches. The low-VOC adhesives are petrochemicals, but they're specially formulated to emit less potentially health-threatening indoor pollution.

Suppliers

WALLPAPERS

Crown Corporation, NA
1801 Wynkoop Street, Denver, CO 80202;
(800) 422-2099 or (303) 292-1313
Anaglypta, a durable wallcovering made of 90 percent recycled cotton and 10 percent wood fiber from Finland.

Euro-Tap
12228 Venice Boulevard #146, Los Angeles, CA 90066; (800) 388-9255 or (213) 398-7033
Wallcoverings made of recycled paper and straw. Eight different textures in white.

Ex
400 East 56th Street, New York, NY 10022;
(212) 758-2593
Cork wallpaper from Sardinia.

Flexi-Wall Systems
P.O. Box 88, Liberty, SC 29657; (800) 843-5394 or (803) 855-0500
Plaster-in-a-Roll gypsum-coated fabric wallcoverings. The Scotland Weave line made of jute and gypsum and available in an array of colors, from Heather White to Highland Thistle, is particularly handsome.

Maya Romanoff Corporation
1730 West Greenleaf, Chicago, IL 60626;
(312) 465-6909
Low-pollutant wallpapers.

Swede-Tech
3081 East Commercial Boulevard, Suite 103, Fort Lauderdale, FL 33308; (305) 771-0204

U.S. distributor is Environmental Construction Outfitters, 44 Crosby Street, New York, NY 10012; (800) 238-5008 or (212) 334-9659 The American importer of glass-fiber wallcoverings made in Sweden. Unlike fiberglass insulation, these wallcoverings are made of longer and larger-diameter glass fibers that are not easily inhaled. They are also specially made to be non-irritating to the skin. They come in a variety of weaves and textures. They're only available in white but are easily painted.

Helen Verrin Custom Prints
available through Environmental Construction Outfitters, 44 Crosby Street, New York, NY 10012; (800) 238-5008 or (212) 334-9659 Custom designed wallcoverings. Available untreated.

WALLPAPER GLUE
AMF Enterprises, Inc.
1140 Stacy Court, Riverside, CA 92507; (714) 781-6860
Low-VOC wallpaper adhesive.

Auro, Sinan Co.
P.O. Box 857, Davis, CA 95617-0857; (916) 753-3104
Natural wallpaper glue.

Devoe and Raynolds Co.
4000 Dupont Circle, Louisville, KY 40207; (502) 897-9861
Natural wallpaper glue.

Eco Design
1365 Rufina Circle, Santa Fe, NM 87501; (800) 621-2591 or (505) 438-3448
Natural wallpaper glue.

Roman Adhesives, Inc.
Calumet City, IL 60409; (708) 891-0188
Natural wallpaper glue.

SEALERS AND FINISHES

A finish is a permanent coating applied to a material to protect it and bring out its beauty. A finish can enhance the natural grain of a wood, for example, and add gloss or sheen. Anything that makes the porous surfaces of a material impermeable is called a sealer finish or, more simply, a sealer. A sealer provides ultimate protection against stains, wear, water, and the weather. Waxes and polishes are not technically finishes. They are "dressings" for finishes, put on for extra protection.

Whether for floors or furniture, siding or masonry, the majority of sealers and finishes on the market these days are either solvent- or water-based petroleum derivatives. Some are formulations that combine natural oils, such as linseed or tung oil, with petroleum distillates. Others are totally synthetic, such as polyurethane, by far the most popular floor finish.

Health and Environmental Concerns

Indoor Air Quality
All petrochemical-based sealers and finishes emit, or offgas, air pollutants called volatile organic compounds, or VOCs (see "A Guide to Home Pollutants," pages 245–249). When used indoors, these products can make the air inside your home unhealthy to breathe. In fact, indoor air-quality experts consider sealers and finishes particularly troublesome because they fall into the category of "wet-applied" materials such as adhesives, paints, and caulks, which are the most polluting products used in home building and decoration.

Air Pollution
The VOCs in sealers and finishes react with sunlight to form ground-level ozone, a major component of smog. VOC emissions are coming under stricter regulation by the federal gov-

ernment and some state governments, most notably California.

Toxicity

Significant quantities of toxic pollutants are produced during the manufacture of many of the binders (resins) and solvents that are the major components of conventional finishes. The pigments and additives used in many of the finishes are also often highly toxic (see "Enamel Paints," pages 185–187). The manufacture of these substances and the toxic wastes produced as by-products of manufacturing are highly regulated by federal and state laws.

Oil Depletion

The resins and solvents that comprise synthetic sealers and finishes are derived from nonrenewable fossil fuels.

Disposal

In some communities, these products and their containers are considered hazardous wastes. In most communities, however, they simply end up in landfills or, more often, are dumped in the sink or down the curb drain.

Environmental Products

Natural finishes are made from ingredients such as plant oils, citrus oil, tree and plant resins, herbal extracts, and beeswax, substances with which humans have lived for hundreds of years. These natural ingredients are biodegradable and are constantly renewed by nature. They contain no known toxic pigments or additives. And they emit no petrochemical-derived volatile organic compounds. However, trees and other plants do emit natural terpenes (or nonpetroleum VOCs), which can pose health hazards to sensitive individuals, particularly in inadequately vented spaces. Of particular concern is the natural solvent d,l-limonene, which the Food and Drug Administration has identified as

a possible irritant (see "Enamel Paints," pages 185–187).

Another option to consider is one of the low-VOC sealers now available. Like conventional products these are petrochemical based, but at least one environmental concern has been addressed: they are specifically formulated to reduce air-polluting emissions.

Use sealers and finishes only when they are essential. And no matter what type of product you choose, make sure the area is well ventilated while you work and for a few days afterward to give air pollutants and odors some time to dissipate.

Wood Siding Finishes

For centuries houses clad with wooden clapboard, shingles, and shakes were left unfinished to weather naturally. Many have lasted for hundreds of years. With two months of exposure to sunlight, wood generally turns yellow or brown, then gray. Dark-colored woods become lighter and light-colored woods become darker. After the initial weathered surface develops, usually in a year or two, further changes occur very slowly. Indeed, only about a quarter inch of softwoods such as pine, cedar, redwood and spruce weathers away every one hundred years.

If you've used a rot-resistant wood on the outside of your house, the best finish from the environmental point of view is no finish at all. However, remaining stands of many of these species, such as redwood and cedar, are in old-growth forests in the Pacific Northwest and it is best to avoid them (see "Woods of Concern," page 101). Moreover, any wood that stays wet is vulnerable to decay. So if you live in a humid area, your house is shaded by tall trees, and the site is poorly ventilated, you'll need to apply a protective finish.

How well your finish performs (and therefore how often you'll need to reapply it) depends a good deal on the quality of the siding (avoid knots and splitting), the density of the wood, how it is cut, its moisture content and surface texture, and how it is installed:

- Heavy woods such as pine or fir shrink and swell more than light woods such as cedar or redwood, putting great stress on the finish.
- Wood siding is best installed and finished when its average moisture content more or less matches that of the area in which it is used: in the arid Southwest, this is about 9 percent; in the rest of the country, about 12 percent. Wood purchased from a lumberyard usually has been kiln dried and has a moisture content of 19 percent or less. It's a good idea to get it to the site as far in advance of construction as possible so that it can acclimate to the site's specific climatic conditions.
- Oil and stains last better on rough-sawn and flat-grained lumber; paints, on smooth, edge-grained lumber and plywoods (see the illustration on page 181).
- Finishes are likely to perform better on vertical siding than on horizontal. One reason for this is because water drains better from vertical siding.
- Weathered wood surfaces provide a very poor substrate for paints and other film-type finishes. Even a few days of exposure will decrease the paintability of your siding. Penetrating finishes are best for even slightly weathered wood.

Conventional Wood Siding Finishes

A number of finishes are widely used on wood siding today:

- Paints. Oil-based paints have been used for decades. They protect siding from water and vapor, but they do not "breathe" and therefore are susceptible to cracking and peeling. Latex paints, particularly the acrylics, are more flexible and hold up better by stretching and shrinking with the wood. Even non-porous oil-based paints are not preservatives, however, and do not prevent decay and rot caused by moisture seeping in from cracks and migrating from the interior of the house. Both oil-based and latex paints are petrochemical products (see "Paints," pages 181–187).
- Opaque Stains. Also called solid color stains, these products are essentially paints. Made with a much higher concentration of pigment than the semitransparent penetrating stains but somewhat less than that of standard paints, they hide or obscure the natural color and grain of the wood. The pigment protects the wood from degradation by the ultraviolet rays in sunlight. Because they form a thin film much like paint, these products are subject to peeling. Like paints, opaque stains are petrochemical products.
- Water-Repellent Preservatives. Water-repellent preservatives are designed to reduce warping, checking, and water stains and to control mildew. They contain mildewcides and a small amount of wax to help repel water, but because they contain no pigment, they provide no protection from ultraviolet rays.
- Water Repellents. These products are simply water-repellent preservatives without the mildewcides. They block moisture but the wood is still susceptible to decay. What's more, like water-repellent preservatives, they do not protect the wood from ultraviolet rays.
- Semitransparent Penetrating Stains. Semitransparent stains penetrate the wood surface to some degree; unlike paint, they do not form a surface film and so do not blister and peel. Some pigment is added to help protect the wood from the degrading effects of sunlight, but not so much that the wood grain is totally obscured. These stains are usually synthetic oil-based. Water-based latex stains are available, but because they don't penetrate

THE ULTIMATE OIL AND WAX FINISH FOR FLOORS

For years, Paul Bierman-Lytle's Masters Corporation experimented with a variety of oils and waxes in a quest for the perfect floor finish. The resulting formula, divulged below, yields a floor finish with unsurpassed depth and sheen. The finish is also extremely long lasting. It is reminiscent of the finishes painstakingly applied by master floor finishers in European homes and palaces of centuries past.

A five-step process is involved: add five parts polymerized tung oil to four parts citrus thinner. Apply two coats of this mixture. Combine equal amounts of polymerized tung oil and citrus thinnner and apply two coats. Melt beeswax in a double boiler and, while still hot, apply in very thin coats with a buffing machine. Reapply the beeswax once or twice a year, depending on traffic.

The polymerized tung oil is available from Sutherland Wells Ltd., P.O. Box 772, Nyack, NY 10960-0772; (914) 348-0100. The company also offers a citrus thinner called DiCitrusol and a natural, nontoxic paste wax. For other suppliers of citrus thinner, see "Paints," pages 181–187. Sources of beeswax are listed in the suppliers list under "Floor Finishes."

---

the wood surface like their oil-based counterparts they are not as effective. Both oil- and latex-based semitransparent stains are petrochemical formulations.

- Transparent Coatings. Clear finishes of spar, polyurethane, or marine varnish are film-forming finishes not recommended for exterior use. Ultraviolet light penetrates the film and the wood degrades quickly, usually within two years.

Environmental Wood Siding Finishes

If you've used rot-resistant woods and your house is located on a sunny site, leave it alone: over time, the wood will attain a natural, weathered patina that no finish can duplicate. If that's not an option, consider a natural, non-petrochemical finish or a low-emission petroleum-based product. The natural finishes are far better than the latter in performance and durability but require more time and care to apply.

If you're painting your siding with one of the natural, non-petroleum paints, treat it first with a wood preservative (see "Wood Preservatives," pages 133–134). Next, apply oil-based primer to create a more even surface and make the paint adhere better. Then apply two coats of oil-based paint: the first coat should be applied within two weeks after the primer; the second, within two weeks of the first coat. Two coats over the primer will last up to ten years. If you apply only one coat over the primer, it will last only half as long (See "Enamel Paints," pages 185–187, for recommended products).

Natural, non-petroleum oil stains are best brushed on. Work in the shade whenever possible, as they will dry more slowly and you'll get fewer lap marks. Apply at least two coats; up to five if you have the time and budget (see "A New House," pages 42–52). One coat will last two to four years, two coats will last up to eight years, and additional coats will last ten years or longer. Apply successive coats before the previous coat dries (if you allow the coats to dry first, additional coats will not penetrate). Use a cloth, sponge, or dry brush, lightly wetted with stain, to wipe off the excess stain that has not penetrated. Always begin with a penetrating primer coat, followed by the pigmented oil stain for UV protection.

Wood Floor Finishes

Until the mid-nineteenth century, Americans lived on bare wood floors. When the floors needed cleaning, they were scrubbed with sand and water. Hard, glossy finishes came into vogue in the Victorian interior. One of the most widespread was shellac, composed of a natural substance secreted by the lac beetle native to India, dissolved in alcohol. Varnishes made with linseed and other natural oils and resins became popular around 1850. Synthetic, petroleum-based resin varnishes were unknown in the pre-1940 house.

Conventional Wood Floor Finishes

Polyurethane accounts for the lion's share of the floor finish market today. An estimated 400,000 gallons of polyurethane are slathered on American floors each year. Polyurethane's popularity stems from the fact that it is quick drying, easy to apply, and provides a tough finish. However, urethane is listed by the Environmental Protection Agency as a proven carcinogen in experimental tests with animals. A variety of other petrochemical-based stains, oil finishes, lacquers, and varnishes are also available.

Like all wood finishes, floor finishes are basically a mixture of resin and solvent. The resin is what leaves a film on your floor, enhancing both its appearance and its durability. The solvent, which is generally the most toxic component of the product, ensures smooth and easy application of the resin.

Environmental Wood Floor Finishes

A good deal of extra time and care is required for application of the all-natural finishes, but your efforts will be well rewarded. They yield a soft, waxlike finish — not the Saran wrap–like, plastic-looking finish produced by polyurethane. They also don't scuff and scratch as easily.

If convenience is a paramount concern, use a low-VOC floor finish. These products are still produced from petroleum, but they've been formulated specifically to lower levels of indoor pollutants.

Woodwork and Furniture Finishes

Unless it was painted, almost all furniture and woodwork was finished with shellac or natural oil-based varnish until synthetic petroleum-based products became the norm. A variety of less-polluting products are now available, including natural resin oils and varnishes, shellacs, and specially formulated, low-VOC synthetics.

Masonry Sealers

A protective sealer should be applied to concrete in basements, and to brick, stone, or any other porous surface used on interior floors, including Mexican tiles, limestone, and terracotta pavers. Either natural oil-based products or low-VOC synthetics are the environmentally sound choices.

Foundation Sealers

Because the foundation of your home is mostly below the surface of the ground, it is vulnerable to water moving through the soil. Health-threatening radon gas, which can seep from the soil through cracks in the foundation and into your house is another concern (see "Radon Prevention Systems," pages 81–90). Foundation walls are typically waterproofed on the outside with a bituminous material such as asphalt. The VOCs emitted by these products, including benzene and napthalene, can be hazardous to installers. A better choice is a product called Dyno Seal. Although it is petrochemical based, Dyno Seal has no odor and is extremely low in VOCs. It also outperforms the competition. Dyno Seal is applied with a brush or roller. It is effective against both water and radon.

Suppliers

WOOD SIDING FINISHES

Natural Finishes

For exterior paints, see "Enamel Paints," pages 185–187

Auro, Sinan Co.

P.O. Box 857, Davis, CA 95617-0857; (916) 753-3104
Natural Resin Oil Glaze — Semi-gloss finish. Available in clear, yellow-brown, medium-brown, and red-brown. For protection against ultraviolet light, clear glaze should be mixed with at least 5 percent Resin Oil Color Concentrates. Use Natural Resin Oil Primer as a primer coat.
Natural Resin Oil Primer — A clear primer.
Natural Resin Oil Color Concentrates — Available in ocher yellow, English red, Persian red, chrome green, ultramarine blue, iron oxide brown, burnt umber, black, and mineral white.

Eco Design

1365 Ruffina Circle, Santa Fe, New Mexico 87501; (800) 621-2591 or (505) 438-3448
Kaldet Resin & Oil Finish — Satin finish. For protection against damage by ultraviolet light, use one of the ready-mixed Kaldet colors (teak, walnut, ebony, rosewood, boxwood, brazil, and weather gray provide UV protection) or add Earthen & Mineral Stain pastes. For best penetration, pre-treat the wood with one or two coats of Dubno Primer Oil.

Dubno Primer Oil — A clear primer.
Linus Linseed Impregnation — A penetrating oil recommended for priming strongly weathered and highly absorbent woods instead of Dubno primer.
Earthen & Mineral Stain Pastes — Available in gold ochre, red ochre, terra di siena, English red, Persian red, rust brown, raw umber, ebony, and oxide green. White and ultramarine blue are not recommended for outdoor use.

Low-VOC Finishes

Kaupert Chemical & Consulting, Inc.

39119 Deerhorn Road, Springfield, OR 97478; (503) 747-2509
Protect-n-Seal — A transparent coating with UV protection.
Seal-n-Oil — A fast-drying, oil-based sealer. Can be applied to damp wood. Available in clear, Sierra redwood, cedar, brown, and gray.

FLOOR FINISHES

Natural Floor Finishes

Natural floor sealers are superior finishes that can be applied alone or in combination with primer oils and finished with beeswax to give your floors a satin sheen. Used alone they will yield a semi-gloss to glossy finish. Two or three coats are required. The primer oils are especially good for porous wood floors because they reduce the number of more expensive final coats required. Apply one to two coats of primer oil. All of the natural finishes will yield a clear finish but will darken the wood slightly. If you want to change the color of the natural wood, apply one of the natural staining and tinting products to the primer oil.

Auro, Sinan Co.

P.O. Box 857, Davis, CA 95617-0857; (916) 753-3104

Natural Resin Floor Oil Sealer — A transparent hardening oil for use on wood and also tile, cork, and stone flooring.

Natural Resin Oil Primer — A clear primer.

Natural Resin Oil Color Concentrates — Available in yellow ochre, Persian red, terra-cotta red, cobalt blue, green, umber brown, earthen brown, earthen black, mineral white, and natural umber.

Floor Wax — A combination of beeswax and plant waxes designed especially for floors.

Eco Design
1365 Ruffina Circle, Santa Fe, New Mexico 87501; (800) 621-2591 or (505) 438-3448
Meldos Hard Sealer — Appropriate for soft woods as well as cork and stone.

Dubno Primer Oil — A clear primer.

Ardvos Floor Oil — A penetrating oil finish especially designed for hardwood floors. Oiled floors can be protected further with Bilo floor wax.

Bilo Floor Wax — The most durable wax in the Eco Design line, suitable for wood and other types of flooring.

Kaldet Resin & Oil Finish can also be used on floors. See "Wood Siding Finishes," above.

Natural Varnishes

Auro, Sinan Co.
P.O. Box 857, Davis, CA 95617-0857; (916) 753-3104
Clear Amber Varnish — Clear varnish with a slight honey tone. The hardest Auro finish for floors.

Eco Design
1365 Ruffina Circle, Santa Fe, New Mexico 87501; (800) 621-2591 or (505) 438-3448
Natural Resin Floor Finish, Satin — For those who like a strong finish with a lot of "body." Recommended for hardwood and softwood floors and cork surfaces. A clear, satin to semi-gloss finish. Can be tinted with pigments.

Natural Resin Floor Finish, Gloss — Essentially the same as the above, but with a high sheen.

Low-VOC Finishes

AFM Enterprises, Inc.
1140 Stacy Court, Riverside, CA 92507; (714) 781-6860
Hard Seal — A medium-gloss floor sealer.

Polyuraseal — A water-based polyurethane that produces a high-gloss finish.

All Purpose Polish and Wax — Can be used to clean and preserve floor finishes.

Henry Flack International
P.O. Box 865110, Plano, TX 75086; (800) 527-4929 or (214) 867-5677
A line of woodworking products that date back to mid-nineteenth-century England, including Mr. Flack's Wood Sealer, composed primarily of secretion of lac beetle and meth spirits. Can be tinted with Mr. Flack's Wood Dye.

Hydrocote
distributed by Terri-Lock in Denver, CO (800) 783-5772; and by Fishbone Enterprises in Norwich, CT, (203) 887-1035.
Hydrocote Clear Polyurethane — Water-based polyurethane finish for floors and other surfaces subject to wear and tear. Can be used both as a sealer and finish coating.

Hydrocote One-Step — Combination stain and polyurethane. Can be used with Hydrocote Clear Polyurethane as a finish coat. Available in natural, golden oak, medium, light and dark walnut, and cherry/mahogany.

Skanvahr Coatings
18646 142nd Avenue NE, Woodinville, WA 98072; (206) 487-1500
Skanvahr EF — A transparent, water-based synthetic finish suitable for wood flooring and

also interior slate, stone, and brick. Can be used with Skanvahr Waterborne Stains.

Skanvahr Woodwax — Can be used over Skanvahr EF for a water-resistant wax floor.

FURNITURE AND WOODWORK

Natural Finishes

Auro, Sinan Co.
P.O. Box 857, Davis, CA 95617-0857; (916) 753-3104
Natural Resin Oil Primer — A clear primer that can be tinted with Natural Resin Oil Color Concentrates.
Natural Resin Lacquers — Clear Glossy Shellac; Clear Velvet Shellac, a semi-gloss finish; and Clear Cembra Shellac, especially effective as a protective inner coating against moths and other insects in cabinets, chests, closets, and the like.
Natural Waxes — Furniture Plant Wax, a water-repellent formulation of beeswax and plant waxes; Larch Resin Furniture Wax, a transparent beeswax paste wax with larchwood resin; Liquid Beeswax Finish, a water-repellent, easy-to-apply liquid.

Eco Design
1365 Ruffina Circle, Santa Fe, New Mexico 87501; (800) 621-2591 or (505) 438-3448
Kaldet Resin & Oil Finish — A water-resistant oil finish. Available in clear and ten colors: pine, teak, walnut, oak, ebony, rosewood, white, boxwood, brazil, and weather gray.
Dubno Primer Oil — Can be used as a primer coat before finishing with Kaldet Resin & Oil.
Tunna Furniture Varnish — Medium- to high-gloss finish. A variety of colors can be obtained by mixing with Earthen & Mineral Stain Pastes.
Earthen & Mineral Stain Pastes — Available in gold ochre, red ochre, terra di siena, English red, Persian red, rust brown, raw umber, ebony, oxide green, white, and ultramarine blue.

Natural Waxes — Laro Antique Wax, a light amber-colored wax especially suitable for antiques and other furniture; Bekos Bee & Resin Ointment, the clearest of the Eco Design waxes that will not discolor light woods; Gleivo Liquid Wax, a cleaner and easy-to-use furniture wax for surfaces that get waxed frequently.
Shellacs — Trebo Shellac, with a light orange cast; Landis Shellac, almost colorless; and Cembra Essence, a natural insect repellent for use in Trebo or Landis Shellacs.

William Zinsser & Co., Inc.
173 Belmont Drive, Somerset, NJ 08875; (908) 469-8100
Bulls Eye Shellac — A product that dates back to 1849, when the original William Zinsser founded the U.S. shellac industry. Zinsser shellacs are formulated with ethyl alcohol solvent.

Low-VOC Finishes

AFM Enterprises, Inc.
1140 Stacy Court, Riverside, CA 92507; (714) 781-6860
Polyuraseal — A low-sheen clear coating suitable for cabinets and woodwork.
Acrylacq — A clear, high-gloss finish.
All Purpose Polish and Wax — Can be used to clean and preserve most surfaces, including furniture.

Henry Flack International
P.O. Box 865110, Plano, TX 75086; (800) 527-4929 or (214) 867-5677
Mr. Flack's Wood Sealer, composed primarily of secretion of lac beetle and meth spirits; Mr. Flack's Wood Dye, water-based stains; Mr. Flack's Liquid Glass, a finishing oil made primarily of vegetable oil and water; and Mr. Flack's French Polish, made of lac beetle secretion and meth spirits.

Hydrocote
distributed by Teri-Lock Distributing in Denver, CO, (800) 783-5772; and by Marc's Restoration in Norwich, CT, (203) 887-1035. Water-based polyurethanes, including Hydrocote Clear Brushable Wood Finish, for paneling, doors, furniture, and cabinets, available in gloss or satin; Hydrocote Clear Wood Finish, for spraying or wiping on; and Hydrocote Clear Polyurethane, for surfaces such as tabletops, counters, and kitchen and baby furniture that require a tough finish.

Skanvahr Coatings
18646 142nd Avenue NE, Woodinville, WA 98072; (206) 487-1500
Skanvahr EF — A transparent, water-based synthetic finish suitable for woodwork and fine furniture, including antiques. Can be used with Skanvahr Waterborne Stains.
Skanvahr Woodwax — Can be used over Skanvahr EF for a water-resistant coating.

MASONRY SEALERS

Natural Sealers

Auro, Sinan Co.
P.O. Box 857, Davis, CA 95617-0857; (916) 753-3104
Natural Resin Floor Oil Sealer — Transparent hardening oil for open-pore materials such as clay tiles, brick, and stones.

Eco Design
1365 Ruffina Circle, Santa Fe, NM 87501; (800) 621-2591 or (505) 438-3448
Meldos Hard Sealer — An oil finish suitable for all absorbent surfaces, including porous stone, terra-cotta tiles, and brick.

Low-VOC Sealers

AFM Enterprises, Inc.
1140 Stacy Court, Riverside, CA 92507; (714) 781-6860
Penetrating Water Seal — Reduces water absorption on porous bricks, Mexican-type pavers, and concrete.
Water Base Mexe Seal — Protects saltillo pavers, adobe, adoquin, grout, granite, concrete, quarry tile, and fired and unfired porcelain from water, stains, and scratches and adds shine.
Water Seal — Forms a tight seal on highly porous surfaces such as grout, tile, and concrete.

Kaupert Chemical & Consulting, Inc.
39119 Deerhorn Road, Springfield, OR 97478; (503) 747-2509
Protect-n-Seal — An all-purpose masonry sealer.

Skanvahr Coatings
18646 142nd Avenue NE, Woodinville, WA 98072; (206) 487-1500
Skanvahr EF — Gives a wet look on stone, enhances the natural color of slate and stone, and makes brickwork easier to clean.

FOUNDATION SEALERS
AFM Enterprises, Inc.
1140 Stacy Court, Riverside, CA 92507; (714) 781-6860
Dyno Seal — A low-VOC synthetic sealer that forms a totally waterproof membrane.

During the 1980s, Americans rediscovered ceramic tiles. Ceramic tile is an ideal material for the environmental home: it's elegant. It's more fire- and heat-proof than vinyl tile. It's almost permanent; while vinyl floors may last five or ten years, ceramic tile will last as long as your house. It's versatile and can be used on floors, bathroom walls, countertops, and backsplashes in kitchens, and in patios, porches, family rooms, and sunspaces. It's also easy to clean with a wipe, resists stains, and shows no signs of aging. The problem is that the tiles are typically installed using toxic mastic and grout. Some European tiles can even contain asbestos.

Conventional Products

Before the chemical revolution, ceramic tile as well as brick, marble, and other stones and glass mosaics were installed over a bed of sand, cement, and water. The cracks between them were filled with a finer mix of the same materials. This method is still used by some craftsmen. However, the majority of installers, and most do-it-yourselfers, use a variety of pre-mixed petrochemical-based mastics, or adhesives, and grouts. To level a floor surface, liquid synthetic latex underlayments are often used. Epoxies are frequently employed as the adhesive to bond the tile to the surface; epoxy systems are basically made up of a synthetic resin and a curing agent, also called a hardener or catalyst. Today, even traditional mastics and cement-based adhesives are often laced with synthetic chemicals to make them stronger or dry faster. Grouts can be cement-based or epoxy-based.

Health and Environmental Concerns

All of the above mastics, epoxies, and grouts can be purchased either as pre-mixed liquids or pastes or in powdered form. The pre-mixed products can be major sources of emissions called volatile organic compounds, or VOCs, that pollute indoor air. They fall under the category of "wet-applied" building materials, which the U.S. Environmental Protection Agency and other indoor air experts consider particularly worrisome. Because they are designed to dry rather quickly, they are formulated with highly volatile solvents and tend to be high in emissions (see "A Guide to Home Pollutants," pages 245–249).

Environmental Products

Traditional, cement-based mastics and grouts that come in powdered form that you mix yourself with water are far more preferable than premixed products because they emit no harmful volatile organic compounds. You should still exercise caution when using them, though, because the dust can irritate your lungs if inhaled, and when mixed with water can injure eyes and skin. The German company Auro offers a natural, non-petroleum-based mastic made from plant resin binders dispersed in water.

Suppliers

AFM Enterprises, Inc.
1140 Stacy Court, Riverside, CA 92507; (714) 781-6860

American Olean Tile
1000 Cannon Road, Lansdale, PA 19446-0271; (215) 855-1111

Auro, Sinan Co.
P.O. Box 857, Davis, CA 95617-0857; (916) 753-3104

Bostik
211 Boston Street, Middleton, MA 01949;
(800) 221-8726 or (508) 777-0100

C-Cure Chemical Company, Inc.
305 Garden Oaks, Houston, TX 77018; (713)
697-2024

H. B. Fuller Company
315 South Hicks Road, Palatine, IL 60067;
(800) 323-7407 or (312) 358-9500

North American Adhesives
530 Industrial Drive, West Chicago, IL 60185;
(800) 637-7753 or (708) 231-7175

When you think of fasteners used in the house, you probably think of nails and perhaps screws. Chances are, glues and adhesives don't come immediately to mind. Yet glues and adhesives are everywhere in the modern home. They're used in cabinetry, furniture, and countertops. They make wall coverings adhere to walls. They're used to glue down carpets and tiles. An entire category of products called construction adhesives are used to bond gypsum wallboard directly to wood and metal framing, moldings to wallboard, paneling to furring strips, subflooring to joists, and so on. To complicate matters, there is a bewildering variety of glues and adhesives, many of them suited only to specific tasks.

Today, most products have a synthetic, petrochemical base. These are called adhesives to distinguish them from the glues of yesteryear, which were made with an animal or plant base, such as hide glues or plant gum glues.

Conventional Products

Some naturally derived glues are still in use. For example, hide glue, made by boiling animal skins, bones, hoofs, and sinews, is the traditional furniture maker's glue, although today it is generally combined with, or replaced entirely by, synthetics. Casein glues, which contain casein, the principle protein in cow's milk, are used to bond woods. Old-fashioned wallpaper paste is derived from plant starches.

However, these days natural glues are far outnumbered by petrochemical-based synthetic adhesives. Some of the most common are:

Acrylic — bonds most surfaces, including wood and masonry
Aliphatic — used for bonding wood
Cyanoacrylate — the instant-set glues, derived from formaldehyde and cyanoacetates

Epoxy — employed on nonporous materials such as ceramic tiles, marble, and masonry

Melamine — commonly used for bonding laminates for doors and countertops

Phenol formaldehyde — the adhesive that bonds exterior-grade plywood and other manufactured wood panels and composite woods together

Polyamide — the hot-melt adhesives used in conjunction with electric glue guns for a variety of purposes

Polyvinyl acetate — all-purpose adhesives and the most popular furniture adhesive

Polyvinyl chloride — all-purpose adhesives

Resorcinol — a formaldehyde-based adhesive used on woods when a waterproof bond is necessary

Styrene butadiene — an adhesive for tiles and other flooring, carpeting, and gypsum wallboard

Urea-formaldehyde — known as plastic resin glue, used for furniture repair and attaching paneling, and in plywoods, particleboards, hardboard, chipboard, and other manufactured wood panels

Urethane — bonds wood, plastic, metal, ceramics, and glass

Health and Environmental Concerns

Synthetic adhesives can be a major source of air pollutants in the home. They fall into a category called "wet-applied" products, which are particularly unhealthy because they contain highly volatile solvents to make them dry quickly. Emissions of formaldehyde and other volatile organic compounds (VOCs) are worst during application. However, the adhesives can continue to offgas pollutants for months (see "A Guide to Home Pollutants," pages 245–249).

Environmental Products

Natural glues are a good alternative to petrochemical adhesives. Be aware, though, that most traditional glues are limited to a particular use and are not weatherproof, unlike the synthetics, which are generally more versatile. The German company Auro makes a line of glues based on new natural, nonpetroleum formulations. For example, Auro's wood flooring adhesive consists of wood resins, plant oils, glues and gums, beeswax, casein, asbestos-free talcum, diatomaceous earth, and water. A handful of firms offer adhesives that are made with petrochemical-based compounds but are specially formulated to minimize offgassing of potentially hazardous pollutants. The emissions of these low-VOC adhesives may be less than one hundredth of those of conventional products. No matter what kind of adhesive you choose, use only as much as you need to get the job done, and make sure the area is adequately ventilated.

Suppliers

NATURAL ADHESIVES

Auro, Sinan Co.
P.O. Box 857, Davis, CA 95617-0857; (916) 753-3104
A line of nine natural adhesives for different uses, such as wood, cork, ceramic tile, linoleum, wall-to-wall carpeting, parquet floors, and wallpaper.

LOW-VOC ADHESIVES

AFM Enterprises, Inc.
1140 Stacy Court, Riverside, CA 92507; (714) 781-6860
Carpet and wallpaper adhesives, 3-in-1 construction adhesive, and Almighty Adhesive, an

all-purpose product for wood, plastic, marble, and ceramic tiles.

Earthbond 7000
600 N. Baldwin Park Boulevard, City of Industry, CA 91749; (818) 369-7371
Flooring adhesive.

Franklin International
2020 Bruck Street, Columbus, OH 43207; (800) 347-GLUE or (614) 443-0241
Titebond ES 747 construction adhesive.

Green Line
available from Environmental Construction Outfitters, 44 Crosby Street, New York, NY 10012; (800) 238-5008 or (212) 334-9659

GL1168 sub-floor and construction adhesive and GL900 drywall and panel adhesive.

North American Adhesives
530 Industrial Drive, West Chicago, IL 60185; (800) 637-7753 or (708) 231-7175
A variety of adhesives for carpeting, wood flooring, and ceramic tiles.

W. F. Taylor Co., Inc.
13660 Excelsior Drive, Santa Fe Springs, CA 90670; (213) 802-1896
The Envirotec Healthguard Adhesives line for carpets and all types of flooring, and for installing corkboard, gypsum wallboard, plywood, paneling, and other rigid panels. For large commercial projects only.

FURNISHINGS

BEDDING

If you want to make your home more environmentally healthy, the bedroom is a good place to start. After all, you spend more time in bed than anywhere else, at least one third of every day and sometimes days on end when you're sick and your immune system needs all the help it can get. More than any other room, the bedroom should be a refuge from pollutants.

Health and Environmental Concerns

Sheets and pillowcases in polyester/cotton blends are treated with formaldehyde to make them wrinkle-free. The consequences of inhaling formaldehyde vapor can range from fatigue to respiratory problems to cancer (see "A Guide to Home Pollutants," pages 245–249). Even 100 percent cotton sheets labeled "no iron" or "easy care" have been treated with formaldehyde. The formaldehyde finish does not come out during washing; indeed, it is designed to last the life of the fabric, although newly manufactured textiles release more polluting vapors than well-worn ones. Mattresses, whether for adults or kids, typically are made with synthetic batting and fabrics and given a liberal dose of petrochemical fire-retardant finish. Pillows are available in polyester fiberfill or synthetic foam.

Blankets also are often made of synthetic fibers, and 100 percent wool blankets in department stores are almost always mothproofed, sometimes with toxic insecticide.

Synthetic bedding not only has potential adverse health effects but is also derived from nonrenewable fossil fuels. Cotton has its own environmental drawbacks: only a small, albeit increasing, percentage of cotton is organically grown. Most cotton is sprayed intensively with pesticides, and chemical defoliants are often used before harvest.

Environmental Products

A totally natural bed, from pillowcase to box spring, can be expensive. But you can significantly lower your exposure to formaldehyde and other noxious vapors by taking a couple of easy and inexpensive steps.

- Avoid sheets and pillowcases made with polyester/cotton blends. Switch to untreated, 100 percent cotton sheets, either percale (smooth) or flannel. Although cotton is not an environmentally perfect material unless it is organically grown, it is healthier than synthetic fabrics. But avoid all-cotton percale sheets labeled "no-iron" or "easy care," which have been treated with formaldehyde.

Don't worry about no-iron cotton flannel because flannel is inherently wrinkle-free and doesn't need a formaldehyde finish. Both cotton percales and flannels are available on sale at department stores and from mail-order suppliers for $50 or less. Linen sheets are an elegant, if more expensive, alternative.

- A natural pillow is the second most important purchase for your bed. Steer clear of synthetics because your nose is right next to your pillow all night, and you don't want to be breathing in any pollutants it may emit. A variety of natural pillows are available, including cotton batting, wool batting, down and/or feathers, kapok (a cotton-like substance found around the seeds of a tropical tree), spun silk, and even buckwheat hulls. Down and feather pillows are sold at just about every department store and the others are available from mail-order suppliers. No matter what kind of pillow you choose, be sure to cover it with an untreated cotton ticking.

- Mattress pads not only protect against stains but to some extent can also block pollutants emitted by petrochemical fibers and finishes in your mattress. Even pads labeled 100 percent cotton are usually filled with polyester batting. A better choice is a mattress pad filled with cotton or wool batting or felt.

- Any comforter or blanket that gets pulled up to your nose on a cold winter's night should be made of untreated natural fibers. Cotton thermal blankets are a good choice in mild weather or used between warmer layers in winter. Nothing is quite as luxurious as a fluffy down comforter; virtually all comforters filled with natural materials are made with 100 percent cotton covers. Wool blankets are made from many different wools and blends, including alpaca, Icelandic wool, llama, mohair, and cashmere. These aren't cheap, but they are very warm and last for many years. Cotton thermals and down comforters are available at department stores. Wool blankets that haven't been moth-proofed can be purchased by mail.

- Replacing a synthetic mattress is expensive because all-natural beds are not mass-produced. They can cost $1,000 or more. The older a synthetic mattress gets, the more pollutants it will have released and the safer it is. Consequently, unless you are allergic to your mattress, it makes sense to hang on to it. When it's time for a replacement, consider a cotton innerspring mattress and box spring or a cotton futon, which is quite comfortable on a wooden-slat bed. Natural, untreated mattresses and futons can be ordered by mail.

Suppliers

Allergy Relief Shop
2932 Middlebrook Park, Knoxville, TN 37921; (615) 522-2795
Low-pollutant bedding.

Bio Clinic, Sunrise Medical
4083 East Airport Drive, Ontario, CA 91761; (800) 347-7780
100 percent cotton covers certified by Scientific Certification Systems.

Bright Future Futons
3120 Central Avenue SE, Albuquerque, NM 87106; (505) 268-9738
Futons.

The Company Store
500 Company Store Road, La Crosse, WI 54601; (800) 356-9367 or (608) 785-1400
Cotton flannel sheets, feather, down, cotton, or silk pillows, down or silk comforters.

The Cotton Place
P.O. Box 59721, Dallas, TX 75229; (800) 451-8866 or (214) 243-4149
Cotton crib sheets, cotton percale sheets and pillowcases, linen sheet sets, cotton felt mattress pads, cotton thermal blankets, cotton bedspreads.

Cuddledown
312 Canco Road, Portland ME 04103; (207) 761-1855
Cotton percale and flannel sheets, down, feather, and cotton pillows, wool-filled mattress pads, down comforters.

Dona Designs
825 Northlake Drive, Richardson, TX 75080; (214) 235-0485
Organically grown cotton pillows, comforters, and futons.

Erlander's Natural Products
P.O. Box 106, Altadena, CA 91003; (818) 797-7004
Cotton percale and flannel sheets and pillowcases, kapok pillows, cotton felt mattress pads, cotton and wool blankets, cotton and wool comforters, cotton bedspreads, cotton futons, cotton mattresses and box springs.

Garnet Hill
262 Main Street, Franconia, NH 03580; (800) 622-6216 or (603) 823-5545
Cotton flannel and linen sheets and pillowcases, down pillows, cotton flannel or wool-filled mattress pads, down comforters, blankets made from fine fibers, including alpaca, llama, merino, lambswool, cashmere, and silk.

Heart of Vermont
The Old Schoolhouse, Route 132, P.O. Box 183, Sharon, VT 05065; (800) 639-4123 or (802) 763-2720
Cotton percale and flannel sheets, wool comforters, cotton and wool blankets, wool pillows, cotton mattresses, cotton futons, wooden-slat beds.

Janice Corporation
198 Route 46, Budd Lake, NJ 07828; (800) JANICES or (201) 691-2979
Cotton percale sheets and pillowcases, cotton pillows, cotton felt mattress pads, cotton thermal blankets, cotton comforters, cotton bedspreads, cotton mattresses and box springs.

Jantz Design
P.O. Box 3071, Santa Rosa, CA 95402; (800) 365-6563 or (707) 823-5834
Natural bedding and wooden beds with natural finishes.

Karen's Non-toxic Products
1839 Dr. Jack Road, Conowingo, MD 21918; (800) KARENS-4
Pillows.

KB Cotton Pillows
P.O. Box 57, DeSoto, TX 75115; (214) 223-7193
Cotton pillows.

Mother Hart's
3300 South Congress Avenue, Unit 21, Boynton Beach, FL 33424; (407) 738-5866
Untreated cotton sheets and towels.

Natural Resources
745 Powderhorn, Monument, CO 80132; (800) USE-FLAX or (719) 488-3630
Natural bedding.

Ogallala Down Comforter Co.
P.O. Box 830, Searle Field, Ogallala, NE 69153; (308) 284-8404
Down comforters.

Richmond Cotton Co.
529 5th Street, Santa Rosa, CA 95401; (800) 992-8924
Natural blankets, children's products.

SDH Enterprises
1717 Solano Way #23, Concord, CA 94520; (415) 685-7035
Untreated bedding and blankets.

Seventh Generation
49 Hercules Drive, Colchester, VT 05446-1672; (800) 456-1177
Untreated bedding and towels.

J. P. Stevens & Co.
1185 Avenue of the Americas, New York, NY 10036; (212) 930-2000
Simply Cotton line of undyed, untreated, and unbleached cotton sheets and pillowcases.

Terra Verde
120 Wooster Street, New York, NY 10012; (212) 925-4533
Natural bedding, including mattresses, futons, and wooden-slat beds. Also, complete natural nursery, including cribs with nontoxic finishes.

Vermont Country Store
Route 100, Weston, VT 05161; (802) 824-3184
Cotton percale and flannel sheets and pillow-cases, down, feather, and wool pillows, wool-filled mattress pads, cotton thermal blankets, wool comforters, cotton bedspreads.

KITCHEN CABINETS AND COUNTERTOPS

Today, a kitchen without built-in cabinets is almost inconceivable. But built-in cabinetry is a relatively new development in the evolution of the American kitchen, dating back only to the 1930s. Before then, pantries and Hoosier cabinets were the primary kitchen storage systems.

Hoosier cabinets were a kind of American variation on the Welsh cupboard. The typical Hoosier cabinet was about forty inches wide and five or six feet tall. It consisted of a deep base cabinet, usually with a door on one side and a column of drawers on the other; a wood or porcelain countertop; and a shallower upper cabinet outfitted with flour sifters, spice racks, and other cooking paraphernalia. The Hoosier cabinet was named after the Hoosier Manufacturing Company, which dominated the field, producing more than 2 million cabinets by 1920. By the late 1930s, most manufacturers of Hoosier cabinets were out of business. These old-time cabinets were replaced by the modular, factory-made models that are ubiquitous today.

Health and Environmental Concerns

Virtually all kitchen cabinets are constructed at least partially of particleboard, although it is used to a somewhat lesser extent in custom and semi-custom versions. The contemporary kitchen cabinet typically consists of a 1/2-inch or 3/8-inch particleboard carcase with a solid hardwood front or "face frame." The shelves are also made of particleboard. The advantage of this material is that it is cheap and provides a smooth and stable substrate for the plastic laminates that are used to cover it. The disadvantage is that particleboard is a major source of potentially health-threatening urea-formaldehyde emissions in the home (see "A Guide to Home Pollutants," pages 245–249). The melamine adhesives used to bond the laminates to

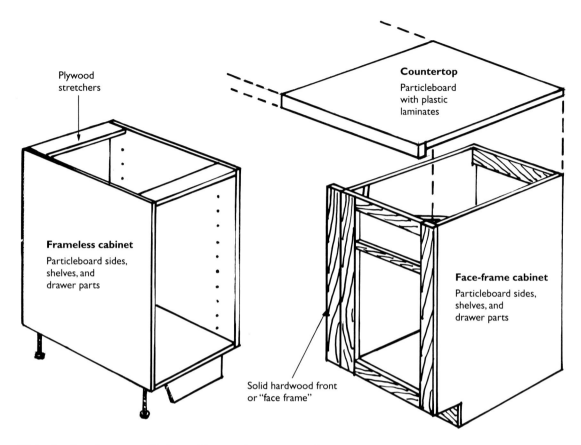

Plywood
stretchers

Countertop
Particleboard
with plastic
laminates

Frameless cabinet
Particleboard sides,
shelves, and
drawer parts

Face-frame cabinet
Particleboard sides,
shelves, and
drawer parts

Solid hardwood front
or "face frame"

Virtually all contemporary kitchen cabinets and countertops are constructed at least partially of particleboard, a major source of potentially health-threatening urea-formaldehyde emissions in the home.

the particleboard can also emit formaldehyde and other VOCs. Most kitchen countertops are made of particleboard with plastic laminates.

Kitchen cabinets raise two additional environmental concerns. Some of the wood species used as veneers, face panels, and trim have been overexploited commercially or are endangered in all or part of their natural range (see "Woods of Concern," page 101). And the finishes used on cabinets and countertops can emit VOCs.

Environmental Products

The core of the environmental kitchen is its cabinetry. This important room should be designed to promote both personal and ecological health,

with sun-dried fruits, nuts, whole-grain crackers, and other tasty treats invitingly displayed. Only materials that are as healthy as they are handsome should be used to build the cabinets in which these wholesome foods are stored. Here are some things to keep in mind when you shop for kitchen cabinets:

- Avoid cabinets made with particleboard. Instead, request cabinets and countertops made with Medite II or other manufactured wood panels bonded with less toxic resins (see "Plywood, Particleboard, and Other Manufactured Wood Panels," pages 120–125). If that's not practical, substitute plywood for the particleboard and paint it with a sealer to encapsulate the formaldehyde fumes. Auro's Clear Cembra shellac or Eco Design's shellac mixed with cembra essence are good choices because not only are they sealers but they also contain cembra, a natural insect repellant (see "Sealers and Finishes," pages 189–197).

- Use woods from certified domestic and tropical forestry operations. Consider less familiar species rather than overused woods such as tropical mahogany or domestic cherry (see "Woods to Look For," pages 100–104).

- Finish your cabinets with a natural, nonpetroleum stain and resin oil or a low-VOC synthetic finish (see "Sealers and Finishes," pages 189–197).

- Look for cabinets that are built to last, with high-quality joinery, hinges, and drawer runners.

- Countertops made of durable natural materials such as granite, marble, or ceramic tile are preferable to plastic laminates. Marble and stone are expensive but inert and toxin free, and they last forever. Ceramic tile can be reasonably priced; be sure to use natural or low-VOC mastic and grout (see "Mastics and Grouts," pages 198–199). Stainless steel cabinetry and countertops are expensive but an excellent choice. Lightweight concrete laid in a steel border is an alternative, too.

- Consider using furniture pieces in the kitchen rather than built-ins. Most kitchens are remodeled many times by each successive occupant. Moveable cabinets can be taken with you when you move, leaving a clean slate for the next owner and cutting down on potential remodeling waste.

Suppliers

Becker Zeyko
1030 Marina Village Parkway, Alameda, CA 94501; (510) 865-1616
Kitchen components handcrafted in Germany's Black Forest.

Canac Cabinets
360 John St., Thornhill, Ontario, Canada L3T 3M9; (800) CANAC-4U
Solid-wood cabinetry.

Crystal Cabinet Works
1100 Crystal, Princeton, MN 55371; (612) 389-4187
Cabinetry constructed without particleboard.

The Masters Corporation
289 Mill Road, New Canaan, CT 06840; (203) 966-3541
Furniture kitchens custom designed, with certified solid wood and Medite, natural, or low-VOC finishes.

Millbrook Custom Kitchens
Route 20, Nassau, NY 12123; (518) 766-3033
Cabinetry constructed without particleboard.

Neff Kitchens
6 Malanie Drive, Brammton, Ontario, Canada L6T 4K9; (800) 268-4527 or (905) 791-7770
Formaldehyde-free kitchen design.

Quality Custom Kitchen
125 Peters Road, P.O. Box 189, New Holland, PA 17557-0189; (717) 656-2721
Cabinetry constructed without particleboard.

St. Charles Manufacturing Co.
1611 E. Main Street, St. Charles, IL 60174; (708) 584-3800
Stainless steel cabinets in a variety of colors.

Woodmode
1 Second Street, Kreamer, PA 17833; (717) 374-2711
Cabinetry constructed without particleboard.

In 1930, chemists Julian Hill and Wallace Hume Carothers developed the first completely synthetic fiber — nylon. Nylon was more prized than natural silk when it was introduced in San Francisco at the 1939 Golden Gate International Exposition. Indeed, the demand for nylon stockings was so great that 64 million pairs were sold the following year, even though nylons cost twice as much as silk. On the heels of nylon came acrylic, polyester, and other synthetic fabrics. Today, synthetics are used around the home in myriad ways, including upholstery, window treatments, sheets, blankets, and table linens.

Health and Environmental Concerns

Synthetic fibers are made of nonrenewable fossil fuels. Even worse than the fibers themselves are the petrochemicals with which they're treated. Almost all fabrics are treated with formaldehyde, for example, which makes them "permanent press." Some of the larger textile manufacturers have provided the federal Environmental Protection Agency with detailed lists of more than two dozen volatile organic chemicals (VOCs) used in the creation of their products — all of which can add to the pollutant levels inside your home (see "A Guide to Home Pollutants," pages 245–249).

Cotton may be natural but is not environmentally benign. In fact, it is one of the most highly sprayed crops in the United States. What's more, petrochemical defoliants are often used to make harvesting the cotton bolls easier.

Environmental Products

The choice of fabrics for the home can be confusing, as a visit to any fabric store or decorator showroom can attest. However, one thing is clear: from oil spills to hazardous manufacturing wastes, the hidden environmental costs of petrochemical-based synthetics are substantial, and so these fabrics are best avoided. Not all natural fibers are problem free, but, unlike synthetics, they do come from abundant, diverse, and renewable sources. It is possible to produce them without an inordinate amount of pollution; even cotton is being produced organically by more and more growers. And processing natural fibers usually requires far less energy than is needed for synthetics.

In the home, different fabrics are suited to different purposes, depending on such factors as durability, strength, and cost. Natural fabrics such as wools and cottons are versatile, strong, and resilient. Linens, although expensive, are very long lasting. Silk, the finest fabric, can be incomparably soft, or it can be stiff, like taffeta. Jutes, hemps, and burlaps can provide a rustic, textured look. Here are some things to consider when using natural fabrics around the home:

- Look for 100 percent natural fabrics, not natural and synthetic blends.

- Use natural fabrics that have not been treated with stain guards and other polluting chemical finishes wherever possible, especially in areas frequented by vulnerable children or the elderly.

- Dyes and mordants are some of the most polluting substances used in textile manufacturing. Unbleached fibers and fabrics that make the most of natural color variations in the fibers — handsome soft shades of off-white, beige, tan, gray, and brown — are the ideal choice. For color, the subtle hues of natural, nonchemical dyes are unsurpassed.

- Don't go overboard with fabric in home decoration. For example, consider solid wood shutters and venetian blinds with natural finishes, bamboo shades, or window treatments that use fabric sparingly and generate a minimum of dust.

Suppliers

The Cotton Place
P.O. Box 59721, Dallas, TX 75229; (800) 451-8866 or (214) 243-4149
Untreated cotton fabrics.

Crocodile Tiers
402 North 99th Street, Mesa, AZ 85207; (602) 373-9823
Untreated fabrics.

Deepa Textiles
333 Bryant Street, Suite 160, San Francisco, CA 94107; (800) 833-3789 or (415) 621-4171
Untreated natural fabrics.

Home Couture, Natural Fabrics
893 S. Lucerne Boulevard, Los Angeles, CA 90005; (213) 936-1302
Naturally dyed fabrics.

Homespun Fabrics and Draperies
Box 3223, Ventura, CA 93006; (805) 642-8111
Ten-foot-wide, textured, untreated cotton fabrics, as well as draperies in two designs.

House of Hemp
P.O. Box 14603, 2111 East Burnside Street, Portland, OR 97204; (503) 232-1128
Hemp fabrics.

Ida Grae
424 LaVerne Avenue, Mill Valley, CA 94941; (415) 388-6101
Naturally dyed natural fabrics.

Janice Corporation
198 Route 46, Budd Lake, NJ 07828; (800) JANICES or (201) 691-2979
Untreated cotton fabrics.

Karen's Non-toxic Products
1839 Dr. Jack Road, Conowingo, MD 21918; (800) KARENS-4
Untreated cotton fabrics.

Jack Lenor Larsen
41 East 11th Street, New York, NY 10003; (212) 674-3993
A variety of untreated cotton and silk fabrics.

Natural Cotton Colours, Inc.
P.O. Box 791, Wasco, CA 93280; (805) 758-3928
Fabrics made from trademarked Fox Fibre cotton that grows naturally in some lovely shades and does not need to be dyed. Much of the cotton is grown organically.

Ocarina Textiles
16 Cliff Street, New London, CT 06320; (203) 437-8189
Natural fabrics.

Organic Interiors
8 College Avenue, Nanuet, NY 10954; (914) 623-2114
Organic cotton fabrics.

Utex Trading Enterprises
111 Peter Street, Toronto, Ontario, Canada M5V 2H1; (716) 282-4887
Silks.

Vermont Country Store
Route 100, Weston, VT 05161; (802) 824-3184
100 percent cottons and cotton damasks.

It makes no sense for us to go to the trouble to build healthy, environmentally sound homes if we then fill them with polluting objects. And we do fill them. The modern American home is out-fitted with armoires, benches, bookcases, buf-fets, hutches, corner cabinets, chairs and stools, chests and dressers, dining tables, entertainment centers, sofas and upholstered chairs, night tables, blanket chests, occasional tables, bed frames, headboards, desks, computer worksta-tions, and more. By dint of sheer quantity alone, our furniture raises environmental questions.

Health and Environmental Concerns

Indoor Air Quality

Studies by the U.S. Environmental Protection Agency showed that when furniture was brought into test homes, indoor formaldehyde levels soared. There are several reasons for this. The synthetic adhesives used to glue furniture parts together can emit formaldehyde and other noxious pollutants called volatile organic com-pounds (VOCs), as can furniture finishes and plastic furniture parts (see "A Guide to Home Pollutants," pages 245–249). The fabrics used on upholstered pieces are routinely treated with stain guards and other VOCs. Polyurethane foam–filled furniture is not only a source of pol-luting emissions but also a serious fire hazard. The foam burns quickly and can fill a room with thick, acrid smoke and toxic gases that of-ten prove fatal.

Resource Depletion

Some of the most prized furniture woods, in-cluding mahogany, teak, rosewood, and ebony, are threatened by overharvesting in all or some of their ranges, and using them adds to destruc-tion of forests (see "Wood Products," pages 97–120). Nonrenewable fossil fuels are used in making plastic furniture parts and synthetic foam cushions and fabrics. The foams may be blown with chlorofluorocarbons (CFCs), which are to be phased out by the year 2000; by their chemical cousins the hydrochlorofluorocarbons (HCFCs), which still damage the protective ozone layer, though to a lesser extent; or by smog-producing pentane gas.

Environmental Products

So what should you look for when you go fur-niture shopping?

- Avoid shelving, tables, and cabinets made with particleboard, which is a major source of indoor urea-formaldehyde fumes. Particle-board, disguised with wood veneer, is almost always used in off-the-shelf stereo cabinets, shelving units, and the like. Make sure the furniture is solid wood. Simple pine pieces are as reasonably priced as particleboard constructions.

- Look for furniture made from certified do-mestic and tropical woods. If you're making your own, seek out less familiar species rather than overused woods (see "Woods to Look For," pages 100–104).

- Furniture that is finished with natural or low-VOC synthetic finishes is best for your fam-ily's health. If necessary, buy unfinished furniture and apply the finishes yourself (see "Enamel Paints," pages 185–187; and "Seal-ers and Finishes," pages 189–197).

- Buy antique and secondhand furniture. These pieces have character and patina. Antiquing is also one of the classiest forms of recycling. If the furniture needs refinishing, use natural or low-pollutant synthetic products.

- Furniture made of bamboo, rattan, and other reeds and grasses and glass, all natural mate-rials in abundant supply, are excellent alter-natives to wood, especially if they have not been finished with VOC-laden chemicals.

- Salvaged woods are sensible substitutes for new wood, especially for outdoor furniture, which ideally is made of redwood, cypress, teak, or other naturally rot-resistant species.
- Recycled-plastic lumber is another good choice for outdoor furniture; see pages 138–140 for suppliers.
- Use safer borate preservatives on outdoor furnishings (see "Wood Preservatives," pages 133–134).
- Upholster your furniture with untreated cottons, wools, rayons, linens, and other natural fabrics (see "Fabrics," pages 208–209).
- Have your furniture made by a local upholsterer using natural jute webbing, natural batting, and down cushions as well as natural fabrics. Suppliers of natural upholstery materials follow.

Suppliers

Bronx 2000
1809 Carter Avenue, Bronx, NY 10457; (718) 731-3931
Big City Forest Products made with wood reclaimed from discarded shipping pallets and crates. Mahogany and other tropical hardwoods, oak, poplar, pine, and fir are combined into a solid butcher block, giving each piece a distinctive look.

Heart of Vermont
The Old Schoolhouse, Route 132, P.O. Box 183, Sharon, VT 05065; (800) 639-4123 or (802) 763-2720
Unfinished furniture made of cherry, maple, oak, and ash responsibly harvested from Vermont's northern hardwood forests and formaldehyde-free glues. Pieces include beds, bedside stands, Mission-style living room furniture, TV stands, and cribs.

The Knoll Group
105 Wooster Street, New York, NY 10012; (212) 343-4167
Maple furniture made from wood certified by Scientific Certification Systems.

The Naturalist
P.O. Box 1431, Provo, UT 84603; (801) 377-5140
Seven collections of furniture, including rustic, western, and contemporary.

Pengelli Forest, Dyfed, Wales, England
Contact Dyfed Wildlife Trust, 7 Market Street, Haverford West SA61 INP, England; 011-44-0437-765-762
Chairs made from thinnings and coppiced timbers. Certified by The Soil Association.

Terra Verde
120 Wooster Street, New York, NY 10012; (212) 925-4533
Custom solid-wood furniture in natural finishes.

The following companies certified by the Rainforest Alliance sell furniture made exclusively from certified tropical woods:

Cooperative Business International
Jalan Karimun Jawa III/1, Klaten, Jawa Tengah, Indonesia; 62-21-272-21077
Fine furniture in mahogany and teak.

The Golden Rabbit
c/o Indo Trade, 22455 Davis Drive, Suite 109, Sterling, VA 22170; (703) 406-8594
Wholesaler of teak furniture and accessories.

Kingsley-Bate, Ltd.
5587B Guinea Road, Fairfax, VA 22032; (703) 978-7200
Teak and mahogany furniture for outdoors.

Latitude 16 Designs
c/o French Harbor Yacht Club, Isla Roatan, Honduras; (504) 45-1460
Wholesaler of garden and patio furniture in lesser-known species.

Smith & Hawken
117 E. Strawberry Drive, Mill Valley, CA 94941; (415) 383-4415
Mail-order retailer of outdoor furniture in Java teak and Honduran cedar.

Victorian Reproductions
P.O. Box 54, La Ceiba, Honduras; (504) 42-0342
Wholesaler of garden and patio furniture and accessories in less familiar species.

The following companies sell Rainforest Alliance–certified Smart Wood products and non-certified tropical wood products. Inquire about the certification status of the products you buy.

Caoba International
3395 Sacramento Street, San Francisco, CA 94118; (415) 292-7809
Manufacturer and retailer of furniture from lesser-known species in Victorian and other styles.

Michael Elkan Studio
22364 North Fork Road, Silverton, OR 97381; (503) 873-3241
Hand-carved mirrors, boxes, and other pieces made from lesser-known tropical woods and temperate woods.

Highland Trading Company
P.O. Box 441, South Royalton, VT 05068; (802) 763-2321
Wholesaler and retailer of compact disc and cassette racks and other products.

The Libra Company, Ltd.
Pentacon Business Estate, Cambridge Road, Linton, Cambridge CB1 6NN, U.K.; 44-0223-893839
Wholesaler and retailer of furniture made from plantation-grown mahogany.

Summit Furniture, Inc.
5 Harris Court, Ryan Ranch Building West, Monterey, CA 93940; (408) 375-7811
Manufacturer and wholesaler of fine teak and mahogany furniture.

The following companies sell natural materials for use in upholstery:

De Ottavio & Sons
127 Van Horn Street, Philadelphia, PA 19123; (215) 627-6344
Untreated cotton batting.

Jantz Design
P.O. Box 3071, Santa Rosa, CA 95402; (800) 365-6563 or (707) 823-8834
Untreated wool batting.

Quaker Jobbing
1722 North Hancock, Philadelphia, PA 19122; (215) 739-9233
Coconut fiber, coconut matting, horsehair.

APPLIANCES AND FIXTURES

APPLIANCES

The typical American house is outfitted with a plethora of appliances that have reduced the drudgery of domestic life. Most kitchens boast a refrigerator/freezer, cooktop, oven (or double oven), dishwasher, microwave, and toaster (or toaster oven), as well as an assortment of portable appliances, including a blender, can opener, food processor, coffeemaker, cappuccino machine, mixer, electric knife, juicer, ad infinitum. Upscale kitchens have two refrigerators, two dishwashers, a warming oven, separate freezer, wine cooler, hot-water dispenser, water filter, bottled water dispenser, garbage compactor, and automatic disposal. In recent years radios, TVs, and computers have also gravitated to the kitchen. And every kitchen has several clocks. The kitchen is by far the most gadget-intensive space in the home. The laundry room or basement also house several big-ticket items, including washer and dryer, de-humidifier, and hot-water heater.

Health and Environmental Concerns

Energy
According to the World Resources Institute, an environmental think tank, home appliances, including water heaters and lights, account for 41 percent of the energy used by the average household, and 51 percent of the cost of household energy. The biggest energy guzzlers are water heaters, refrigerators, and freezers.

Ozone Depletion
The refrigerant used in refrigerators and freezers is HCFC-22, better known as Freon. HCFCs are related to CFCs, the family of chemicals that has been linked to ozone depletion in the upper atmosphere. CFCs are very stable and do not break down easily when released into the air; some can remain in the atmosphere for a century or more, destroying ozone. HCFCs contain hydrogen and therefore break up more quickly and have a shorter life span, but they, too, deplete ozone to some extent. By international agreement, CFC production must be totally eliminated by the year 2000. Right now, HCFCs are slated for elimination by the year 2030, but this date may be accelerated because recent studies suggest that HCFCs may be more destructive of ozone than was originally believed.

Noise
The din produced by exhaust fans, refrigerator compressors, dishwashers, disposals, compactors, alarms, and the myriad of portable appli-

ances make the kitchen and laundry room two of the noisiest spaces in the house.

Environmental Products

Appliances of the twenty-first century will be mercifully quiet. They won't require Freon or other ozone-depleting chemicals. They'll be designed for optimum recyclability, and be made of materials with the fewest environmental impacts; stainless steel appears to meet these criteria better than aluminum or plastics. Appliances of the future will not only conserve water and energy but will be powered by renewable energy sources such as photovoltaics or solar electricity.

We're already making some headway: every year American appliances grow more energy efficient. And some dishwashers and washing machines are designed specifically to conserve water.

The National Appliance Energy Conservation Act of 1987 established minimum efficiency standards for major home appliances. In 1988, for instance, all new dishwashers were required to have a cold-water-rinse option, saving energy that would otherwise be required to heat the water. Progressively stringent standards for refrigerators took effect in 1990 and 1993. According to the Washington, D.C.–based nonprofit group American Council for an Energy-Efficient Economy (ACE[3]), if you replaced a typical 1973, 18-cubic-foot refrigerator with a more energy-efficient 1990 model, you've saved about 1,160 kilowatt hours a year ever since. This translates into a savings of $116 a year on your electric bill, assuming your utility charges $.10 per kWh (many charge more). It also saves over a ton of emissions per year of carbon dioxide, one of the major global warming gases.

Although the new standards eliminate the worst energy wasters in showrooms, refrigerators, dishwashers, and other appliances are still available in a wide range of energy efficiencies,

so it still pays to shop carefully. Don't go shopping without a copy of *Consumer Guide to Home Energy Savings,* an inexpensive little book published and updated regularly by ACE[3]. (For information on how to obtain a copy, write to the American Council for an Energy-Efficient Economy, 1001 Connecticut Avenue, NW, Suite 535, Washington, DC 20036.) It's packed with buying tips, including lists of the most efficient heating and cooling systems, water heaters, refrigerators and freezers, cooktops and ovens, dishwashers, washing machines, dryers, and lights available. It does not include European models, which often are very energy and water conserving; nor does it list solar-electric appliances.

Some highly efficient appliances cost the same as energy guzzlers. Others cost more initially but will yield savings in the long term. When you buy a refrigerator, for example, not only do you pay the sales price, you also commit yourself to paying the costs of running it over its lifetime. This can add up: running a refrigerator for fifteen to twenty years typically costs about three times what it costs to buy it in the first place. Buying an energy-efficient model will save you money over time, often a substantial amount. The true cost of purchasing and running an appliance is called its life-cycle cost, a key piece of information that the "EnergyGuide" labels in showrooms lack. *Consumer Guide to Home Energy Savings* tells you how to calculate life-cycle figures so you can compare the true costs of various models.

Another critical consideration when you're shopping for a new refrigerator or freezer is whether or not it is HCFC-free. The Sun Frost company in Arcata, California, which has been manufacturing and selling some of the world's most energy-efficient refrigerators for years, now offers HCFC-free models. The Whirlpool Corp. was named the winner of a $30 million competition among mainstream American appliance manufacturers, sponsored by the

Environmental Protection Agency, to produce an HCFC-free, super-efficient refrigerator. Whirlpool's alternative refrigerant is HFC-134-A. Unlike CFCs and HCFCs, HFCs contain no chlorine and do less damage to the ozone layer. The company's new super-efficient, non-HCFC model is a 22-cubic-feet, side-by-side refrigerator/freezer measuring 33 inches wide, 32 inches deep, and 66 inches high. It is almost 30 percent more energy efficient than 1993 models. Whirlpool expects to manufacture about 250,000 of these refrigerators between 1994 and 1997. The $30 million award will be used to offer rebates to buyers. Several European manufacturers, including the German company AEG, also make HCFC-free models that are being test marketed in Europe but are not yet available in this country.

For more information, see "Heating and Cooling," pages 145–154, "Hot-Water Heating," pages 216–218, and "Lighting," pages 226–232.

Suppliers

AEG, Andi-Co Appliances, Inc.
65 Campus Plaza, Edison, NJ 08837; (800) 344-0043
Energy- and water-efficient stainless steel appliances.

ASKO
(800) 367-2444
Energy-efficient dishwashers, washers, and dryers made in Sweden.

Bosch
2800 South 25th Avenue, Broadview, IL 60153; (800) 866-2022 or (708) 865-5200
German-made, quiet, energy-efficient, and water-conserving dishwashers.

Electron Connection
P.O. Box 442, Medford, OR 97501; (916) 475-3401
Appliances.

Energy Concepts Co.
627 Ridgely Avenue, Annapolis, MD 21401; (301) 266-6521
Solar thermal refrigerator.

Frigidaire
(800) 451-7007
Refrigerators that use 50 percent fewer HCFCs than typical models and rate 30 percent above federal standards for energy efficiency.

Miele Appliances, Inc.
22D Worlds Fair Drive, Somerset, NJ 08873; (800) 289-MIELE or (908) 560-0899
Energy- and water-conserving stainless steel appliances.

Norcold
600 South Kuther Road, Sidney, OH 45365; (800) 752-8654
Gas-powered refrigerator.

Regency USA Appliances, Ltd.
P.O. Box 3341, Tustin, CA 92680; (714) 544-3530
Gas convection wall oven. Circulating hot air, fan-forced. Cleans while cooking. Smokeless flame grill. Outside power venting.

Sunelco
P.O. Box 1499, Hamilton, MT 59840; (406) 363-6924
Appliances.

Sun Frost
P.O. Box 1101, Arcata, CA 95521; (707) 822-9095
The most energy-efficient American-made refrigerators. HCFC-free models now available.

The company makes units that can run on conventional 110-volt alternating current or on 12- or 24-volt direct current for hookup to photovoltaic systems. One drawback is that Sun Frosts are made largely of plastic components.

Traulsen & Co., Inc.
114-02 15th Avenue, College Point, NY 11356; (718) 463-9000
Stainless steel refrigerators. Not energy efficient and use HCFCs. However, the stainless steel is a big environmental plus, and the compressor can be located outside the home, eliminating noise indoors.

Whirlpool Corp.
CAC Center Pipestone, 2303 Pipestone Road, P.O. Box 0120, Benton Harbor, MI 49022-2400; (800) 253-1301

HOT-WATER HEATING

When you consider all the ways we use hot water — showers, baths, clothes washers, dishwashers, pools, hot tubs, spas — it's not surprising that the domestic hot-water heater is one of the two biggest energy hogs in the house. The typical hot-water heater consists of a tank in which water is heated with natural gas, oil, electricity, or, in remote locations, propane. Swimming pools are usually heated with propane or electricity or else are tied into the larger domestic hot-water heating system.

Health and Environmental Concerns

The biggest environmental problem associated with the domestic hot-water heater is the amount of energy it uses. Eighteen percent of all energy used in the American home goes to heating water. Multiply that by the millions of systems in the country and you're talking about not only staggering amounts of nonrenewable fossil fuels, but also the full range of environmental impacts associated with its production and use — including emissions of carbon dioxide, one of the major global warming gases (see "Heating and Cooling," pages 145–154).

Environmental Products

A ground-source system, which provides clean, efficient space heating and air conditioning as well as all the domestic hot water you need, is the environmental first choice for heating water (see page 149). However, if you already have another kind of heating system, a solar flat-plate heater is the next best choice. Over their lifetimes, these hot-water systems will save on fossil fuels used by conventional water heaters and drastically reduce the production of air pollutants that cause acid rain and global warming.

In the 1970s, solar water heaters were hot

commodities. Federal tax credits during that decade of uncertain energy supplies and soaring prices made them attractive investments. But the solar hot-water heating industry went bust in the early 1980s. Poorly designed hardware and badly trained installers stole some of the glow from solar systems, but the biggest factor contributing to the bust was the loss of federal as well as many state tax credits. Even without tax write-offs, solar-heated water is still a cost-effective option for most of the United States, even New England, where the summers are short and the winters harsh. Indeed, solar hot-water heating is not just for environmental activists and technology enthusiasts anymore — about 1.2 million households in this country already use them. We're still light years behind Japan, where there are approximately 1.5 million solar hot-water systems in Tokyo alone. However, practically every home can be retrofitted with a solar hot-water system. Best of all, today's solar water heaters are much better looking, better built, and more reliable.

Solar hot-water systems are technologically simple. The two main components are a solar collector and a water storage tank. The most commonly used collector is the copper flat-plate. It consists of a black copper absorber plate, insulation, and a glass cover. As the sun shines, radiation passes through the glass and strikes the absorber plate, where it is converted to heat. The copper absorber plate contains a series of vertical and horizontal tubes for carrying heated fluid. Gone are the days when large, ugly collectors littered rooftops. The new versions are slim and lie supine at the same angle as the roof; indeed, an increasing number are actually being built into the roof, so that all you see is the sleek, silver-blue glint of the glazing, with the hardware tucked discreetly out of view.

The various kinds of solar heaters available in the 1970s have been distilled down to two basic types: the closed-loop antifreeze system and the drainback system. Both designs avoid the Achilles' heel of many early solar water heaters: winter freeze damage to the pipes and tanks. Both are so-called active systems because they use pumps to move the fluid around. The most advanced systems now use a small solar-electric-powered pump to circulate the fluid, making the system 100 percent sun powered.

The closed-loop antifreeze system, used in northern climates, utilizes a freeze-proof solution in a loop between the collector panel and the heat exchanger in the hot-water storage tank. In other words, the water you shower with never comes in contact with the copper collector panel. The drainback system, used in southern climates, utilizes water as the medium flowing through the collectors. However, as soon as the temperature inside the collectors (which normally is much higher than the outside temperature when the sun is shining) drops below about 40 degrees F, the pump stops and the water in the collector drains back into a special holding tank. When the collector warms up again, the system refills.

The advantage of using water is that it is capable of holding more heat than antifreeze, and so the system operates more efficiently. The advantage of using antifreeze is extra freeze protection.

Other than the type of technology, there are three primary concerns to bear in mind when planning a solar hot-water system: the size of the panel(s) and storage tank, the angle and orientation of the panels, and the panels' exposure to sunlight. The square footage of the panels is directly linked to how much hot water you use. If you install a low-flow showerhead and faucet aerators and water-conserving washer and dishwasher, you'll need a smaller — and less expensive — array. The size of the panels also varies greatly from region to region and even within different microclimates in the same region. The collectors must be oriented south for good exposure to sunlight. And they should be mounted at the proper angle to the sun (your dealer can help you figure out how many square feet of solar collector you need and the optimum angle for installation). Ideally, the panels should be completely unshaded from 9 A.M. to 3 P.M. The only houses that are not suitable for solar hot-water heaters, therefore, are those with southern obstructions, such as buildings or large stands of evergreen trees.

A properly sized system should supply 70 to 80 percent of your hot-water needs. In most areas of the country, a backup heating source is necessary, which can be natural gas, oil, electricity, or propane.

Cost-wise, solar hot-water heaters are ever more attractive. Prices are dropping, and some systems now retail for about $2,000, or $3,000 with installation. That's more expensive than a conventional system. But because ordinary hot-water heaters cost so much to run, solar systems can pay for themselves quickly. For example, in some areas it can cost $500 a year to operate an electric hot-water boiler. This means that a solar water heater can be paid off in as little as four or five years — and after that, it is *saving* you a lot of money.

Some innovative forms of financing are making solar hot-water heating even more attractive. Some utilities offer rebates; others, low-interest financing. In addition, mortgages guaranteed by the Federal Housing Administration cover an industry-certified solar hot-water system, meaning the cost can be amortized over 20 or 30 years. It may add $15 or $20 a month to your mortgage payments, but you'll be saving $30 to $40 a month in energy costs.

A solar heating system is by far the most economical way to heat a pool. Solar costs a little as one fifth as much as a household hot-water system for this task. Solar collectors for swimming pools usually need not be glazed, saving money. The unglazed panel can be located on a sunny slope below the pool or on the roof, or it can be integrated into a handsome poolside arbor.

Suppliers

Solar Energy Industries Association
122 C Street NW, 4th Floor, Washington, DC 20001; (202) 383-2600
They can tell you if the collector of the solar hot-water system you're eyeing has been cleared by the Solar Rating and Certification Corp., an entity consisting of state officials and industry representatives.

Florida Solar Energy Center (FSEC)
300 State Road 401, Cape Canaveral, FL 32920-4099; (407) 783-0300
An even better rating system for solar hot-water systems is called the OG-300 test, done by FSEC. It tests the complete system rather than just the collector. The FSEC can provide you with extensive information about various solar hot-water systems.

The Europeans used the term "water closet" for the toilet, and the term is apt. Conventional American commodes use five gallons per flush, accounting for about 28 percent of water consumption inside the home. This kind of water waste is difficult to justify at a time when more and more communities are facing water shortages. What's more, the costs of supplying high-quality water, expanding or renovating pipes and plants, and treating contaminated sources increase every year. So does the cost of treating the enormous volumes of wastewater that result from our inefficient use of this crucial resource.

Environmental Products

Water saved by efficient appliances and fixtures can be used to stretch existing reservoir supplies so that shortages are less likely during drought years. In fact, conserving water can eliminate the need to build costly new reservoirs or rely on impure sources of drinking water. It also enables watershed managers to maintain water levels in streams high enough to meet the needs of fish and other wildlife.

In the past ten or fifteen years, a new generation of environmental commodes that use only 1.6 gallons per flush has been refined. In these ultra-low-flush toilets, the tank size and working mechanism and the shape and fluid dynamics of the porcelain bowl and goose-necked water trap have all been redesigned for water conservation.

According to the Rocky Mountain Institute, an environmental think tank, replacing a conventional toilet with an ultra-low-flush toilet can reduce household water consumption by 21.6 to 59.4 gallons per day, which translates to 7,880 to 21,700 gallons per year — a savings of 58 to 78 percent. Ultra-low-flush toilets now come in enough shapes and designer colors to satisfy the most discerning buyer. And they cost as little as $100.

Ultra-low-flush toilets make good financial, as well as environmental, sense. Remember, the toilet accounts for more than a quarter of indoor water use, and installing an ultra-low-flush model can slash this amount by half or three quarters or more. The Massachusetts Water Resources Authority estimates that retrofitting a 1.6-gallons-per-flush toilet will save a Boston homeowner $74 per year in 1989 dollars. This means an average payback on investment in only three years, or an annual return on investment of 39 percent — making the toilet a better investment than a savings account and most stocks! What's more, if you're building a new home, the savings accrued due to a reduced septic leach field can easily exceed the added cost of not only ultra-low-flush toilets but also low-flow showerheads and faucet aerators.

A toilet can last the lifetime of a house. Consequently, it's a good idea to spend some time evaluating which model is best for you and your family. Here are some things to keep in mind:

- Just because a toilet is called a "water miser" or "water saver" by the manufacturer doesn't mean it is a 1.6-gallons-per-flush ultra-low-flush model.

- There should be no more operating problems with a good ultra-low-flush toilet than with a conventional model. However, just as you do when you buy a car, you should do your homework, visit the showroom, shop around, and test the model you're thinking of buying.

- Don't go for the absolute lowest gallons per flush, unless water in your area is very scarce. Most toilets that use less than one gallon per flush are mechanical models that depend upon pumps, compressors, and other mechanisms. They don't work for long when the power fails, and any extra hardware you add is just another potential headache and could require costly repairs. Also, any toilet with less than a 1.6-gallon flush may require more frequent cleaning with a toilet brush. None of the toilets listed below are mechanical models. Most use 1.6 gallons per flush; a few use between 1.1 and 1.6 gallons per flush.

Suppliers

American Standard
P.O. Box 6820, Piscataway, NJ 08855; (800) 752-6292 or (201) 980-3000
New Cadet Aquameter, Elderly New Cadet Aquameter, Phebe Aquameter.

Artesian Industries
201 East 5th Street, Mansfield, OH 44901; (419) 522-4211
Santa Fe.

Briggs Plumbingware
4350 West Cypress Street, Suite 800, Tampa, FL 33607; (800) 627-4447 or (813) 878-0178
Ultra Conserver.

Crane Plumbing
1235 Hartrey Avenue, Evanston, IL 60202; (708) 864-9777
Cranemizer.

Eljer Industries
901 10th Street, Plano, TX 75074; (800) 4ELJER2 or (214) 881-7177
The Contoura, The Preserver I, The Preserver II, The Patriot, The Terrace, Triangle Ultra.

Gerber Plumbing Fixtures Corporation
4656 West Touhy Avenue, Chicago, IL 60646; (312) 675-6570
Aqua Saver.

Kohler Co.
444 Highland Drive, Kohler, WI 53044; (800) 456-4537 or (414) 457-4441
Couture Lite, Portrait Lite, Wellworth Lite.

Mansfield
150 East 1st Street, Perrysville, OH 44864; (419) 938-5211
Allegro, IFO Cascade, Quantum.

Peerless Pottery
P.O. Box 145, Rockport, IN 47635-0145; (800) 457-5785 or (812) 649-6430
Hydromiser.

Porcher Inc.
3618 E. LaSalle, Phoenix, AZ 85040; (800) 359-3261 or (602) 470-1005
Veneto Water Closet.

Titon Industries
P.O. Box 566848, Atlanta, GA 31156; (404) 399-5252
Flush-Lite, Euro Flushlite.

Universal Rundle Corporation
217 North Mill Street, New Castle, PA 16103; (412) 658-6631
Atlas, Hercules, Saturn, Taurus.

Water Conservation Systems, Inc.
9 Pond Lane, Concord, MA 01742; (508) 369-3951
Sovereign.

Western Pottery
11911 Industrial Avenue, Southgate, CA 90280; (213) 636-8124
Aris.

LOW-FLOW SHOWERHEADS AND FAUCET AERATORS

The environmental bathroom is outfitted not only with an ultra-low-flush toilet but also a low-flow showerhead and faucet aerator. Faucet aerators are also useful in the kitchen.

The conventional showerhead gushes five to eight gallons of water per minute. Although it can vary from household to household, shower water use can account for 21 percent of total indoor water use as well as roughly 33 to 40 percent of total hot-water use, according to the Rocky Mountain Institute, an environmental research group. Consequently, there is a large potential for water and energy savings. The latest low-flow models with high-velocity pulsing jets use only one to two and a half gallons a minute. It's important to note that low-flow showerheads are not the flow restrictors (washers with a small hole in the middle) used in some communities in times of water shortage. These devices do save water but also produce a pathetic trickle, resulting in a not-very-satisfying shower.

The products recommended below are not flow restrictors. Although they use no more than 2.5 gallons per minute, they offer the same invigorating torrent to which we've become accustomed. They vary in "feel," from "misty," "needle-like," and "pulsating" to "vigorously pounding" flows. Because a shower is a very personal experience, it's a good idea to test some low-flow models before buying. Incidentally, you needn't give up your hand-held showerhead to conserve water; a number of low-flow models are now available.

Most low-flow showerheads cost less than $20. If you were to invest the same amount in a certificate of deposit, it would take about 10 years to return $20 in interest, while a savings account would take close to 20 years. In 10 years, your efficient showerhead, on the other hand, will easily return 10 to 40 times its cost in saved energy alone, not counting the value of the saved water.

Faucet aerators are inexpensive devices that add air to the water to provide the same pressure with less water flow. In the bathroom, where performance doesn't matter much, go with the lowest-flow model, 1.5 gallon per minute or less, at the cheapest price. In the kitchen, because you'll be filling pots and pans, you'll need a higher flow, but 2.0 to 2.5 gallons per minute is generally sufficient. Because they are so inexpensive and widely available at local hardware or plumbing-supply stores, it's not necessary to list suppliers here.

Installing your new showerheads and aerators is simple enough to do between halves of a basketball game; all you need is a lockjaw wrench, pipe wrench, or wide-mouthed adjustable pliers. Keep in mind, though, that if your existing shower gets uncomfortably hot when someone flushes the toilet, then a low-flush showerhead may well exaggerate the problem. However, this is easily solved by having your plumber install an antiscald device.

Suppliers

Chatham Brass Company, Inc.
5 Olsen Avenue, Edison, NJ 08820; (800) 526-7553 or (908) 494-7107
Chatham Adjustable Spray showerheads, Chatham Flow Control showerheads.

Energy Technology Labs, Inc.
2351 Tenaya Drive, Modesto, CA 95354; (800) 344-3242 or (209) 529-3546
Shower-Spa 2000, Shower-Spas Next Generation 2001.

Interbath, Inc.
665 North Baldwin Park Boulevard, City of Industry, CA 91746; (800) 423-9485 or (800) 828-7943 in California or (818) 369-1841
Ondine Super Watersaver.

Melard Mfg. Corporation
153 Linden Street, Passaic, NJ 07055; (201) 472-8888
Melard Water-Saving Showerheads.

Moen Incorporated
25300 Al Moen Drive, North Olmsted, OH 44070; (216) 962-2000
Easy Clean, Easy Clean Deluxe, Moenflo, Moenflo Deluxe, Pulsation.

Niagara Conservation Corporation
45 Horse Hill Road, Cedar Knolls, NJ 07927; (800) 831-8383 or (201) 829-0800
Niagara line of showerheads and various other water-saving products.

Omni Products, Chronomite Laboratories, Inc.
21011 South Figueroa Street, Carson, CA 90745; (800) 447-4963 or (800) 447-4962 in California or (310) 320-9452
Omni Energy Saver line of showerheads.

Pacific Environmental Marketing & Development Co.
421 South California Street #D, San Gabriel, CA 91776; (800) 243-8775 or (818) 292-3855
The Rainshow'r Dechlorinator, low-flow showerhead that removes 90 percent or more of the chlorine from shower water.

Resources Conservation, Inc.
P.O. Box 71, Greenwich, CT 06836-0071; (800) 243-2862 or (203) 964-0600
The Incredible Head, The Incredible Head Europa line.

Whedon Products, Inc.
21A Andover Drive, West Hartford, CT 06110; (800) 541-2184 or (203) 953-7606
Whedon SaverShower line of showerheads.

If booming sales of bottled waters and filtration equipment are any indication, Americans are worried about the quality of their water. Lead can leach into drinking water from home plumbing. Natural hazards such as radon in water have recently come under the scrutiny of the federal Environmental Protection Agency. And water supplies, public and private alike, are beginning to show the effects of two hundred years of industrialization. But it is possible to make sure that the water you use for drinking, cooking, and bathing is safe.

Health and Environmental Concerns

Pesticides, industrial chemicals, and other man-made pollutants are at the top of most people's list of concerns. These synthetic toxins are indeed present at high levels in domestic water in some areas. More immediate and widespread dangers, however, are posed by lead, radon, and nitrates.

Lead

Studies have shown that significant levels of lead in drinking water are far more common than was once assumed. What's more, lead levels once considered safe are now known to threaten health, particularly the health of fetuses, infants, and children. Acute lead poisoning can cause severe brain damage and even death. The effects of chronic, low-level exposure are more subtle; exposure at a young age can cause permanent learning disabilities and hyperactive behavior. Low-level lead exposure has also been linked to elevated blood pressure, chronic anemia, and nerve damage.

Lead gets in water primarily from corrosion of plumbing that contains lead. Many houses built from about 1919 to 1940 have service pipes made of lead, and even newer homes in colder regions may have lead pipes. Lead can also leach out of lead solder, which was banned in 1986 on pipes that carry drinking water.

Radon

According to one EPA official, waterborne radon may cause more cancer deaths than all other drinking-water contaminants combined. The agency estimates that at least 8 million Americans may have undesirably high radon levels in their water supply. Showering, dishwashing, and laundering all agitate water and can release radon into the air.

If your drinking water comes from rivers, lakes, or reservoirs you have little to worry about; radon is most likely to be present in groundwater in an area considered to be a potential hot spot for the radioactive gas. If your indoor radon level is high and you use groundwater, you should definitely test your water. For information on how to test your water, when to take action, and recommended filtration devices, see "Radon Prevention Systems," pages 81–90.

Nitrate

Nitrate contamination, like radon contamination, occurs mainly in groundwater. Most at risk are infants who drink formula mixed with nitrate-rich water. Bacteria in infants' digestive tracts convert nitrate to nitrite, which in turn combines with hemoglobin in the blood to form a compound called methemoglobin, which cannot transport oxygen. The resulting condition, methemoglobinemia, deprives vital organs of oxygen. Although rare, methemoglobinemia can cause brain damage or death. Some adults, including pregnant women, may also be susceptible.

High nitrate levels are most common in agricultural areas. Chemical fertilizers and manure from feed lots are especially rich sources of nitrogen compounds, which are converted to nitrate in the soil. The nitrate migrates down through the soil to underground water. Lawn fertilizers and leaky septic systems have caused elevated nitrate levels in suburban areas as well. High nitrate levels may be a sign that other pollutants, including pesticides and bacteria from septic tanks, are also present.

Organic Chemicals

Organic pollutants are petrochemical compounds containing carbon, including industrial solvents such as trichloroethylene (TCE), pesticides, and trihalomethanes (THMs). While most organic chemicals pose only localized problems, THMs, which are by-products of the chlorination of water, a measure taken by most municipalities to prevent waterborne disease, are widespread. Because there is some evidence that THMs contribute slightly to cancer risk, the EPA requires water-supply systems serving more than ten thousand people to keep levels below one hundred parts per billion. Chloroform is another worrisome by-product of chlorination. The agency also regulates or recommends maximum levels of scores of other organic chemicals, some suspected cancer causers.

Environmental Products

Getting your water tested is the first line of defense. If your water comes from a public water system, ask the utility or the appropriate government agency for a copy of its latest water analysis. Federal law requires most public water companies to test the water regularly and make the results available to the public. If you don't get satisfactory answers, or if your water comes from a private well, have your water tested independently by an EPA-certified lab that can give you their certification number when you speak to them. Here are some things to look for:

- Test public water for high levels of chloroform and trihalomethanes.

- If your water comes from a private well, have it tested periodically for bacteria, inorganic compounds, and radon. If you live within a mile or two of a gasoline station, refinery, chemical plant, military base, or landfill, test for organic chemicals. If you live in an agricultural area, test for nitrate and pesticides.

- No matter where your water comes from, have it tested for lead if your house is more than thirty years old or if the pipes are joined with lead solder.

If testing turns up problems, water purification equipment can effectively remove many contaminants. But it's important to note that you must buy a filter designed to remedy your particular problem or problems; different kinds of water filters are best at removing specific types of pollutants. Remember, too, that water filters require regular maintenance if they are to remain effective.

In addition to the different kinds of filters, there are whole-house units that remove pollutants as they enter the home and point-of-use units that are attached to the kitchen faucet or showerhead. Whole-house filters are desirable because they remove pollutants not only from tapwater but also from water used for showers, baths, swimming pools, dishwashers, and clothes washers. They must be used to remedy some problems, such as radon. Point-of-use filters are smaller and more affordable. They are the filters of choice for removal of lead at the tap. Stainless steel or glass filters are preferable to plastic canisters.

There are three major types of water filters:

Carbon Filters

Carbon, in the form of granulated activated charcoal or carbon block, filters are the best for extracting radon, trihalomethanes, organic chemicals, and pesticides from water. Carbon block filters generally trap more pollutants than granular-carbon units. High-volume carbon filters are most effective; don't rely on a faucet-mount or pour-through filter. Avoid carbon filters that are impregnated with silver nitrate, a poison and suspected carcinogen. Your carbon filter will last a lot longer if you install a sediment filter to remove solids before they reach the carbon and clog it prematurely.

Reverse Osmosis

In reverse-osmosis systems, pressure in the water line pushes the water against a semipermeable membrane. Large molecules are trapped; water and small organic molecules pass through. Reverse osmosis is best at removing salt, nitrate, lead, and other heavy metals. It also has significant drawbacks: the reverse-osmosis filters available for home use make limited amounts of water for drinking or cooking. Most need three to six hours to process one gallon. They also waste water — only 10 to 25 percent of the water passing through the unit is forced through the membrane. The rest goes down the drain.

Distillers

Distillers boil water, then cool the steam until it condenses; the resulting distillate drips into a jug. Salts, sediment, and heavy metals such as lead, which won't boil or evaporate, stay behind in the boiling pot. However, many organic chemicals can pass through a distiller and end up in the distilled water. What's more, distillation is a slow process; it can take five hours or more to make a single gallon of water. Like reverse-osmosis units, they waste a lot of water. Distillers have the additional disadvantage of requiring a considerable amount of energy to convert water into steam.

Absolute Environmental's Allergy Products and Services Store
2615 South University Drive, Davie, FL 33328; (800) 329-3773 or (305) 472-3773
Activated carbon filters.

Culligan International Co.
1 Culligan Parkway, Northbrook, IL 60062; (708) 205-6000
Reverse-osmosis filtration system.

Environmental Purification Systems
P.O. Box 191, Concord, CA 94522; (800) 829-2129 or (510) 682-7231
Omni shower filter installed on showerhead to remove chlorine. Recommended for those sensitive to chlorine.

FILTRX Corporation
11 Hansen Avenue, New City, NY 10956; (914) 638-9708
A complete line of filters for lead, scale, bacteria, and chlorine.

General Ecology, Inc.
151 Sheree Boulevard, Lionville, PA 19353; (610) 363-7900
Seagull carbon filtration systems.

Global Environmental Technologies
P.O. Box 8839, Allentown, PA 18105; (215) 821-4901
KDF water-filtration system.

Kentrel Corporation
P.O. Box 173, Avoca, PA 18641; (800) 437-9200 or (717) 451-0622
Kentrel solid carbon block filtration systems.

Kiss International

1475 12th Street East, Palmetto, FL 34221;
(813) 722-3999
Reverse-osmosis filtration systems.

Lowry Aeration Systems

4915 Prospectus Drive, Suite C1, Durham, NC
27713; (919) 544-9080
The Stripper aeration system designed to reduce
radon levels in the household water supply to
less than 200 picocuries per liter.

Multi-Pure Drinking Water Systems

P.O. Box 368, Barre, MA 01005; (800) 735-
6542
Carbon block filtration system.

Pacific Environmental Marketing & Development Co.

421 South California Street #D, San Gabriel,
CA 91776; (800) 243-8775 or (818) 292-3855
On-tap filtration and dechlorination systems.

Pure Water Place

P.O. Box 6715, Longmont, CO 80501; (303)
776-0056
Seagull IV four-step purification system.

Real Goods

966 Mazzoni Street, Ukiah, CA 95482-3471;
(800) 762-7325
Mail-order supplier of The Rainshower filter installed on the showerhead to remove chlorine.
Recommended for those sensitive to chlorine.

LIGHTING

More information reaches the human brain
through the eyes than through any other sense
organ. It's no wonder, then, that lighting has
been a primary concern in human shelters since
prehistoric peoples used fire for night vision.
Early New Englanders stuck pieces of resinous
wood between the stones of their fireplaces to
impart bright, although smoky, flames to light
their homes. Candles were made from splinters
of wood dipped in tallow, beeswax, or whale
oil, called spermaceti; indeed, candlepower, the
standard unit of measurement of light intensity,
is based on the amount of light emanating from
one spermaceti candle. The arrival of gas lights
in Victorian homes vastly improved the quality
and convenience of home lighting. In the twentieth century, electric lighting became available
at the flick of a switch. Unfortunately, this lux-
ury is dependent entirely on the consumption of
nonrenewable fossil fuels to make the electricity — and has resulted in massive air pollution.

Conventional Products

The common incandescent light bulb hasn't
changed much since Thomas Edison invented it
more than a century ago. Edison's bulb used
electric current to heat a filament until it
glowed. Today's incandescents consist of a
tungsten wire in a glass bulb filled with inert
gas, usually argon. Light is produced when
enough electric current passes through the wire.

Incandescent bulbs come in a variety of
shapes and wattages and sizes. You can get
them at the supermarket. Everybody knows
how to screw one into a socket. They're even
used in cartoons to signify a bright idea. We're
so used to them, in fact, that they are the standard by which other light sources are compared.

A GUIDE TO HOME LIGHTING

The wide array of new lighting options is enough to make your head spin. Here's a quick guide to the types of bulbs, tubes, and fixtures generally best suited to your various home lighting needs.

OUTDOORS

Spotlights and floodlights	Halogen
Landscape lighting	Low-voltage halogen
Tennis and basketball courts	High-intensity discharge
Swimming pool	Low-voltage halogen

INDOORS

General lighting	Compact fluorescent
Task lighting	Compact fluorescent, low-voltage halogen
Workshop	Fluorescent
Utility lighting	Compact fluorescent
Bathrooms, kitchens, greenhouse	Full-color-spectrum

Health and Environmental Concerns

Lighting and Health

Human beings, like virtually all other living things on the planet, have evolved with natural light, and our bodies are adapted to its daily and seasonal cycles. The most obvious evidence of this is that we normally sleep at night and wake with the morning light. Exposure to the ultraviolet rays in sunlight is essential for the formation of vitamin D and healthy bones. Intuitively we all know that light affects our moods. In 1980, Dr. Alfred Lewy, then at the National Institute of Mental Health, conducted studies that demonstrated that bright light affects the pineal gland's production of the hormone melatonin. As a result, we are closer to understanding the condition called seasonal affective disorder (SAD), a kind of depression as-sociated with the short, low-light days of winter in many parts of the world. Subsequent research has shown that not only the intensity and duration of light but also the presence or absence of certain wavelengths in the electromagnetic spectrum can have effects on our bodies.

Today few of us live in harmony with the natural cycles of the sun. We barely live with sun at all, given the fact that we spend more than 80 percent of our lives indoors. This wouldn't be so bad if our homes and offices were designed to make the most of natural light. They're not. We depend on artificial lighting not only at night but also to a considerable extent during the day. And most types of artificial light differ in important ways from natural light — for example, in the range of colors and invisible wavelengths they produce. Headaches, poor vision, seasonal affective disorder, and other kinds of depression are some of the better known health effects of our overreliance on conventional artificial lights. New findings indicate that artificial lighting can have health ramifications far beyond those imagined by science even a decade ago. One particular concern is our continual exposure to the electromagnetic fields in our homes (see "A Guide to Home Pollutants," pages 245–249).

Energy Efficiency

When enough electricity passes through the tungsten wire in a common incandescent bulb to make it glow, it produces light, and heat. In fact, most of the energy goes to producing heat. For this reason, incandescents are by far the least energy-efficient form of lighting available today. The new compact fluorescent lamps, for instance, consume 75 to 85 percent less electricity than their incandescent counterparts to produce the same quantity of light.

Multiply the amount of energy wasted by each incandescent bulb by the millions of such bulbs in use across the country and the total energy waste is truly staggering. In the United States lighting consumes about a quarter of all electricity generated — about 20 percent directly, plus another 5 percent in cooling equipment to compensate for all the heat our lights emit. The Lawrence Berkeley Laboratory calculates that about 80 or 90 percent of the electricity used for lighting could be saved if we converted to the most efficient equipment in our homes and businesses. More than half of the electricity we use on lighting could be saved cost effectively. Halving our consumption of electricity would have very real environmental benefits: cutting in half the amount of oil, coal, and uranium used in power plants would result in dramatic reductions in acid rain–producing sulfur dioxide emissions as well as carbon dioxide emissions, which scientists have linked with potential global warming, fewer strip mines, less radioactive waste, less imported oil, and fewer oil spills, not to mention less consumption of precious fossil fuels.

Environmental Products

Daylight

For all the reasons mentioned above, daylight is a crucial feature of the environmental home. Whether you're designing a new home or making improvements on an existing structure, there are ways to increase natural light. Design the home first with daylight in mind. Build daylight into a brand-new structure by putting bedrooms on the east side to face the sunrise, breakfast rooms to take advantage of the early-morning sun, and kitchens, living rooms, and sunspaces on the south side of the house. If you're remodeling, you can add large windows that flood the room with natural light, or install skylights. Be sure, though, to have properly sized roof overhangs to shade your windows from the hot summer sun, and use one of the new window technologies that feature maximum R-values to prevent heat loss in winter in cool climates and low shading coefficients to prevent overheating in warm ones (see "Windows," pages 141–144).

Proper Placement

Once you've done all you can to make the most of natural light, take a hardheaded look at where you need supplemental lighting and for what purpose.

Outside, you may want strategically placed lights for security. Consider, too, some fixtures that enable you to find your way up a driveway or walkway or light up an entry to the house. You may also need additional lights to illuminate the swimming pool or tennis court or show your landscape to best effect. Special energy-efficient lamps are available for these purposes.

Inside, there are a number of different kinds of lighting needs. General lighting, usually a central overhead fitting, is useful in hallways or vestibules. Task lighting is ideal for areas where you read, work, and cook. Mood lighting can be especially pleasing to accent an antique, say, or show off artwork. Garages and closets can get away with functional utility lighting. Full-color-spectrum lights, which come closest to replicating natural daylight, are best in rooms where true color rendition is important — the kitchen, the bathroom, and the greenhouse, for example.

Alternative Lighting

A number of alternatives to the common incandescent bulb are available.

Halogen — Tungsten halogen lights are a newer, better version of the common incandescent bulb. They consist of a filament encased in a halogen gas, usually iodine or bromine. This improves performance in two ways: first, the filament burns hotter and brighter, increasing efficiency; new halogen bulbs containing krypton gas are more efficient still. Second, tungsten is recycled back to the filament, not deposited on the bulb wall, where it blocks some of the light; this not only maintains the intensity of the light but also increases the life of the bulb. Halogens last three times longer than standard incandescent bulbs of equal wattage and burn 10 percent brighter. These bulbs are particularly useful as spotlights and floodlights.

Low-Voltage Halogen — These 12-volt lamps are even more energy-efficient than regular halogens. They come with a transformer that enables them to run on standard 110-volt house current. They're designed for desks, drafting tables, reading chairs, and other places where you do close work. They're also useful for illuminating the swimming pool and landscaping.

Fluorescent — Although fluorescent lights have been around for a while, they're worth listing as an environmental alternative to incandescents because they're vastly more energy efficient and produce significantly less heat, saving on cooling costs. Fluorescent bulbs are available in straight, U-shaped, and circular designs. Light is produced when phosphor coatings on the inner tube wall are activated by ultraviolet radiation generated by a mercury vapor arc. A ballast is required to strike the arc and maintain proper current and voltage when the light is on. Fluorescent lights have gotten better and better in recent years. Old-fashioned fluorescents produced a harsh, cold light that gave skin tones a blue or green appearance. Manufacturers found that by tinkering with the coating on the inside of the tube they could almost duplicate the warm quality of incandescents, which emphasize reds and yellows. Redesigned electronic ballasts also make it possible for the bulbs to come on instantly with no flicker, and no noise or hum. Nothing beats traditional tube-type fluorescents for bright, even lighting in a work area.

Compact Fluorescent — Compact fluorescent lights are small fluorescent tubes bent into compact shapes. They come in a variety of sizes as well as shapes, and some are only slightly larger than conventional incandescents. A special adaptor allows the bulb to be screwed into existing household light fixtures, from table lamps to track lights to sconces; some models include adaptor and bulb in one. If you're replacing old incandescent fixtures, you can get ones that are specially made for compact fluorescent bulbs. Like the larger versions, compact fluorescents consume as much as 75 to 85 percent less energy than incandescents and last more than ten times longer. Like halogens and larger fluorescents, they cost more, too. But if you balance the higher initial cost against the reduction in replacement costs, you can recover the money you spent buying the bulb, and more. You can make money without even factoring in the savings in electricity. In the words of energy expert Amory Lovins of the Rocky Mountain Institute, "This is not a free lunch. This is a lunch you are paid to eat." Some electric utilities offer rebates to customers who buy compact fluorescent bulbs. Ask your power company if they do, too.

High-Intensity Discharge — This lighting category actually includes three types of lamps: mercury, metal halide, and high-pressure sodium. High-intensity discharge lights all utilize sealed arc tubes in which extremely bright light is produced by a high-pressure discharge of gas. Because they produce such intense light, they are used primarily outdoors. The metal halide lamps are most appropriate for residences, to light up the tennis or basketball court, for example. They're far more energy efficient than even fluorescents, and they last about as long or longer. Both metal halide and high-pressure sodium lamps are used indoors as grow lights for plants.

Full-Color-Spectrum — These light sources are designed to closely replicate the spectral balance of daylight. They are available both as incandescent bulbs and fluorescent tubes.

Electromagnetic Field–Shielded — These are specially designed to shield you from the electromagnetic radiation produced by electrical wiring and appliances, including lights. Right now, only one manufacturer is producing EMF-shielded bulbs and fixtures. It is also possible to protect yourself and your family from electromagnetic fields by having your electrical wiring installed properly (see "A Twenty-first Century Home," pages 67–76).

Determine which type of bulb (technically called "lamp") or fixture is most suitable for each of your lighting needs, then select the most energy-efficient option available. At first glance it would seem that converting to an "all compact fluorescent" home is the way to go. This will save energy, but it won't give you the highest-quality lighting. A combination of all types of energy-efficient lamps is recommended (see "A Guide to Home Lighting," page 227).

Lighting Control Systems
A number of devices from simple to high-tech can make the lighting systems you choose even more energy efficient:

Dimmers — By allowing you to lower the intensity of light emanating from a fixture, dimmers offer additional opportunities to conserve energy. Generally, dimmers are most appropriate where lighting needs vary widely. For example, lights near windows can be dimmed to very low levels during the day, then gradually turned up as night falls. They also enable you to set the mood of a space. (Be aware, though, that most compact fluorescents are incompatible with dimmers.)

Timers — Timers have been around for many years. They've been employed mostly for security to turn lights on and off when no one is home. However, they can also help you save energy. They're most useful in areas in which you follow a fairly set routine. For example, you can set a timer to automatically turn on outdoor lights at dusk in winter, when you and your family most likely get home from work or school after dark, and turn them off at an hour when the whole family has typically returned home.

Motion Sensors — Unlike timers, occupancy sensors, also known as motion sensors, are a relatively new technology. They include a sensing device and a low-voltage relay to switch lights on or off in response to the presence or absence of people. A variety of different sensors are available, including passive infrared, which detect body heat, and ultrasonic, which detect soundwave disturbances caused by motion. Unlike timers, motion sensors are most useful in areas where you come and go intermittently — bathrooms, for example.

Light Sensors — Light-sensing controls measure ambient lighting levels with a photocell and activate or deactivate fixtures accordingly. These have been widely used for outdoor lighting but hold great promise for indoors, particularly to control lights near windows or skylights. They are especially effective when used in conjunction with dimming controls; as ambient light increases or decreases, the lighting fixture can be turned up or down in increments.

Manual Programmable Lighting System — These systems divide the house into zones in which every domestic electrical system, including lighting, can be controlled. The main objective is to give you room-by-room control. If you're in the TV room, for instance, and you want to turn up the heat and turn off the lights, you don't have to go very far to do it. The systems have up to seven "programs." If you want several lights to come on in the evening for security when you are on vacation, for example, you can program the system to respond accordingly. The systems can be reprogrammed as frequently as you wish.

Computer Management Systems — These more sophisticated systems include hidden sensors in each room, which send messages back to a central computer. The computer, as programmed by you, responds to the message by, for example, turning on lights when you turn up the driveway or walk down a hall. Again, the main objective is to save energy. Many of the early computer-management systems were extremely complicated, but newer versions are more user friendly. They're still expensive, but prices are coming down.

Suppliers

Real Goods
966 Mazzoni Street, Ukiah, CA 95482-3471; (800) 762-7325
A mail-order supplier of a variety of environmentally sound lamps and fixtures.

Seventh Generation
49 Hercules Drive, Colchester, VT 05446-1672; (800) 456-1177
Environmental lighting products by mail order.

COMPACT FLUORESCENT LIGHTS
North American Philips Information and Literature Center
114 Mayfield Avenue, Edison, NJ 08837; (800) 631-1259

Panasonic Industrial Company
Two Panasonic Way, Secaucus, NJ 07094; (800) 899-1199 or (201) 348-5380

FULL-COLOR-SPECTRUM LIGHTS
Environmental Light Concepts, Inc.
3923 Coconut Palm Drive, Suite 101, Tampa, FL 33619; (800) 842-8848

EMF-SHIELDED BULBS, FLUORESCENT TUBES, TASK LIGHTS, AND FIXTURES
Ott-Lite, distributed by Environmental Construction Outfitters
44 Crosby Street, New York, NY 10012; (800) 238-5008 or (212) 334-9659

Litetouch

3550 South 700 West, Salt Lake City, UT 84119; (801) 268-8668

Crestron

101 Broadway, Cresskill, NJ 07626; (800) 237-2041 or (201) 894-0660

See suppliers under "Energy Management Systems," pages 164–165.

Natural Lighting Company

7021 W. Augusta, Suite 106, Glendale, AZ 85303; (800) 960-LITE or (602) 435-6542 Passive Daylighting by NLC daylighting systems, Active Daylighting by NLC daylighting systems, and So-Dark motorized shade screens.

CHILDREN'S PRODUCTS

Because children spend a good deal of the day at home, most parents go to great lengths to make the house safe. They childproof everything in sight to prevent cuts, electrocution, and poisoning. But not many parents are aware that the typical home is full of hazards that can cause chronic problems — from asthma to neurological damage to cancer.

Yet children, especially the very young, are more vulnerable than adults to many environmental toxins. Children have a higher metabolic rate, which means that they require more oxygen and therefore breathe in two to three times as much air — and air pollutants — relative to body size as adults. They're also more active physically, which increases their breathing rate and intake of pollutants even more. Children consume more water relative to body weight than adults — about twice as much, in fact. They also absorb some water pollutants at a higher rate than adults; for example, children absorb about 50 percent of the lead they eat and breathe, compared with 10 percent for adults. Children are at greater risk simply because their growth and development may make them more susceptible to toxic substances. For instance, the human nervous system develops very rapidly for several years after birth and is not completely mature until adolescence. During this long period of maturation, the developing brain may be particularly sensitive to toxins. The young are very vulnerable to carcinogens because children's cells are still dividing rapidly during infancy and early childhood, resulting in a greater probability that a genetic mutation will occur that initiates the cancer process.

To make matters worse, children have greater exposure to home pollutants than adults do. They play close to the floor, where many pollutants settle. They cram anything and everything into their mouths. Outdoors, they dig holes in the ground, roll around on the grass, hide in the bushes, and climb trees, coming into contact with pesticides on plants and in the soil.

Nevertheless, most of the laws and regulations designed to protect us from the daily assault by pollutants have been set with adults in mind. The standards for toxic chemicals permitted in the air, in drinking water, and in the food supply, for example, are all based upon adult exposure or consumption patterns, systematically ignoring the special risks that toxic substances pose for the very young.

Health and Environmental Concerns

The prevalence of childhood asthma is on the increase. The symptoms of asthma are triggered by exposure to allergens and air pollutants, including tobacco smoke, house dust, paints, and household cleaning products. Air pollution affects children's health in other ways; poor-quality air has reduced the lung capacity of children in Los Angeles, for example, to, on average, 10 to 15 percent less than children who live in less polluted areas. And no matter where they live, children, who spend much of their time inside the home, are constantly exposed to the various indoor pollutants: carbon monoxide, nitrogen dioxide, and sulfur dioxide from the burning of fossil fuels in kerosene and gas space heaters, gas stoves, wood stoves, fireplaces, and central heating systems; passive cigarette smoke; formaldehyde in particleboard, bedding, fabrics, and countless other home products; 4-PC in synthetic carpeting; and electromagnetic fields from power lines, electrical wiring, and appliances, to name a few.

Cancer among children 14 and younger in the United States increased 21.5 percent between 1950 and 1986, according to the National Cancer Institute. Because children tend to spend more time at home and in basement playrooms than adults, their exposure to radioactive radon gas is higher. A potential cause of childhood leukemia, recent epidemiologic studies and the occurrence of suspicious "clusters" of cancer cases suggest, are the electromagnetic fields emitted by electrical wires and appliances. Nitrate poisoning from polluted drinking water is almost exclusively a problem that afflicts children. In their immature digestive systems, the nitrates are converted to nitrites, which interfere with the ability of blood to carry oxygen, causing a condition called methemoglobinemia; poisoned babies develop blue lips and skin and in severe cases may experience brain damage and death.

Surpassing all of these potential hazards is the number one environmental threat facing children — lead. An estimated 3 to 4 million children in the United States have enough lead in their bodies to lower IQs, impair learning, and cause nerve damage and kidney disease.

You can dramatically reduce your children's exposure to dangerous pollutants around the home. The following checklist highlights what you can do about the worst hazards, both indoors and outside. For more information, see "A Guide to Home Pollutants," pages 245–249, and the sections on specific materials and alternative products.

Indoors

- Have your water tested for lead, nitrates, radon, pesticides, and other chemical pollutants. If necessary, install the appropriate water-filtration equipment (see "Water Filters," pages 223–226).

- Test radon levels inside your home, particularly if you live in an area at high risk for the radioactive gas. If levels are high, take corrective action (see "Radon Prevention Systems," pages 81–90).

- Conduct a comprehensive survey of all potential pollutants inside your home, including asbestos, lead paint, tobacco smoke, sources of formaldehyde, and other potentially toxic volatile organic compounds (VOCs). Call in professionals for asbestos and lead abatement.

- Reduce your children's exposure to dangerous combustion gases and toxic air pollutants. Choose natural or low-VOC synthetic products for all the interior surfaces of your home, especially those in their bedrooms and playrooms (see "Paints," pages 181–187, "Wallpaper," pages 187–189, "Carpets and Underlayments," pages 170–174, "Resilient Flooring," pages 175–178, and "Sealers and Finishes," pages 189–197). Make sure your home is ventilated properly (see "Mechanical Ventilation," pages 154–157).

- Prohibit smoking in your home.

- Avoid furniture, particularly children's furniture, constructed with urea-formaldehyde-laden particleboard (see "Furniture," pages 210–212).

- Use safer, low-VOC cleaning and maintenance products (see "Home Maintenance Products," pages 238–244). Store even these products out of the reach of children.

- Don't buy toys made of petroleum-based plastics and synthetic rubber to limit exposure to VOCs. Look for toys made of renewable, natural materials such as wood, cork, untreated cotton, and wool.

- Avoid wall-to-wall carpeting, which harbors dust, dust mites, mold, and other allergens and is a "sink" for chemical pollutants. Synthetic carpeting and backings themselves are also a major source of VOCs. Use area rugs made of natural fibers instead.

- Install a HEPA (high-efficiency particulate absolute) filter in your kids' bedrooms if you don't have a whole-house unit hooked into your central air system (see "Air Filtration and Purification," pages 157–159).

- Use only natural, untreated and formaldehyde-free bedding (see "Bedding," pages 202–205).

- Reduce your children's exposure to electromagnetic fields. Use EMF-shielded computer and television screens. Avoid electric blankets. Keep children's beds away from cables that bring electricity into the home. (See "EMF-Shielded Wiring and Appliances," page 161, for more on managing electromagnetic fields.)

- Make a point of using only healthy, resource-conserving products in the home, and encourage your children to participate in discussions of the environmental merits of their toys and the materials used to furnish and decorate their rooms.

Outdoors

- Don't use synthetic pesticides and weed killers in your yard. They are not only a health hazard to children as well as adults, but also kill the natural predators of the pests you're trying to eliminate. Begin an integrated pest management program, using the least-toxic approach to any particular problem (see page 240).

- Avoid pressure-treated wood, for children's playsets and outdoor furniture in particular. Instead, use naturally rot-resistant woods that don't require toxic preservatives or coatings (see "Rot-Resistant Woods," page 104).

- When building outdoor structures for children, use materials that don't release VOCs, such as lightweight concrete, metals (except aluminum, which is extremely energy intensive to produce), earth, and stone in addition to wood.

- Explore salvage yards for other imaginative materials, including old boats, doors, ropes, wooden wheels, concrete, or metal sculptures and the like.

- Keep play areas away from electric power lines.

- Integrate natural landscapes, not just boring lawn, in play areas — with, for example, water sculptures, a pocket prairie or meadow filled with interesting native grasses and wildflowers that will attract butterflies and other wildlife, or a patch of native forest, complete with trees to climb and perhaps a treehouse. Design and build the playground with the site itself rather than imposing pre-manufactured pieces onto the site. Most play areas today are pre-created stage sets, leaving little to a child's imagination. Many of the most inspiring playgrounds mimic "survival" or "obstacle" courses, complete with rope-web climbing trellises, ropes to climb up to towers with slides to get back down to the ground, and other challenges.

Suppliers

Hanna Anderson
1010 NW Flanders, Portland, OR 97209;
(800) 222-0544
Clothing.

Biobottoms
Box 6009, Petaluma, CA 94953; (707) 778-
7945
Clothing.

The Blue Earth
2899 Agoura Road, Suite 625, Westlake
Village, CA 91361; (800) 825-4540 or (818)
707-2187
Baby care products.

The Cotton Place
P.O. Box 59721, Dallas, TX 75229; (800) 451-
8866 or (214) 243-4149
Untreated cotton crib sheets.

Earth's Best, Inc.
Box 887, Pond Lake, Middlebury, VT 05753;
(802) 388-7974
Organic baby food.

Eco Design
1365 Rufina Circle, Santa Fe, NM 87501;
(505) 438-3448
Children's art materials.

HearthSong
Box B, Sebastopol, CA 95473; (800) 325-2502
Art supplies.

Heart of Vermont
The Old Schoolhouse, Route 132, P.O. Box
183, Sharon, VT 05065; (800) 639-4123 or
(802) 763-2720
Natural bedding for children, sleeping bags,
furniture, toys.

Hugg-A-Planet
247 Rockingstone Avenue, Larchmont, NY
10538; (914) 833-0200
Toys.

Karen's Nontoxic Products
1839 Dr. Jack Road, Conowingo, MD 21918;
(800) KARENS-4
Art supplies.

Krafty Kids, Inc.
11358 Aurora Avenue, Des Moines, IA 50322-
7907; (515) 276-8325
Craft supplies.

The Masters Corporation
289 Mill Road, P.O. Box 514, New Canaan,
CT 06840; (203) 966-3541
Custom designs for playgrounds, fantasy structures, children's furniture, children's bedrooms
and bathrooms.

The Natural Baby Co.
RD 1, Box 160S, Titusville, NJ 08560; (609)
737-2895
Clothing.

Pastorini Spielzeug of Zurich
Industriestrasse 4, Postfach, 8600 Dubendorf,
Switzerland; (01) 821-55-22
Wood toys, games, crafts, water games.

Real Goods
966 Mazzoni Street, Ukiah, CA 95482-3471;
(800) 762-7325
Toys and games.

Seventh Generation
49 Hercules Drive, Colchester, VT 05446-1672; (800) 456-1177
Toys, clothing.

Terra Verde
120 Wooster Street, New York, NY 10012; (212) 925-4533
Complete natural nursery.

Waterforms, Inc.
Route 177, P.O. Box 930, Blue Hill, ME 04614; (207) 374-2384
Designs, manufactures, and installs water sculptures and flowing water environments for interior and exterior settings.

HOME
MAINTENANCE
PRODUCTS

In a nation where using spotted drinking glasses is tantamount to social sin, it's no wonder that home maintenance products are a big environmental problem. In fact, they're some of the most hazardous substances found around the house. They come in many permutations: all-purpose cleaners, oven cleaners, rug cleaners, glass cleaners, toilet bowl cleaners, metal polishes, furniture polishes, spot removers, flea powder, laundry detergents, bleaches, upholstery cleaners, mold and mildew removers, drain decloggers, disinfectants, paint strippers, scouring powders, bug sprays, and on and on. To make matters worse, many of these products are kept in the kitchen or pantry where food is prepared or stored and the risk of contamination is high. Often they're stored under the sink, in easy reach of children, or in the basement near the furnace or boiler, increasing the risk of explosion or fire.

Safer and less environmentally damaging alternatives are inexpensive — so much so, in fact, that replacing hazardous cleaning products and pesticides is an easy way to begin transforming your house into a healthy, ecologically sensitive place to live.

Health and Environmental Concerns

Cleaners are the most toxic chemicals used around the home. They're also the only household products regulated by the federal Consumer Product Safety Commission that are not required to have ingredients listed on the label. The CSPC does require that consumers be warned on the label in relative terms how dangerous a product is. The classifications, in order from most toxic to least toxic, are: "Toxic or highly toxic," a product that is poisonous if eaten, inhaled, or absorbed through the skin; "Extremely flammable," "flammable," or "combustible," a product that catches fire easily; "Corrosive," a product that eats away your skin (such as oven cleaners, which contain lye); "Irritant," a product that causes skin rashes and/or inflammation of mucous membranes; and "Strong sensitizer," a product that may provoke an allergic reaction.

What toxins, specifically, are in household cleaners? Some information is available from poison control centers. For example, some furniture and floor polishes contain phenol, which when ingested in even small amounts may cause nausea, vomiting, convulsions, coma, respiratory failure, and death. Paint strippers may contain methylene chloride, a mild narcotic. Effects from intoxication include headache, irritability, and numbness in the limbs. In severe cases, the chemical has caused hallucinations, coma, and death.

Home pesticides, which are regulated by the U.S. Environmental Protection Agency, are required to have the active ingredient listed on the label, as well as "signal words" to warn users of their relative toxicity. Pesticides labeled "Danger" or "Poison" are highly toxic; those labeled "Warning" are moderately toxic; and the ones labeled "Caution" are slightly toxic. Acute or immediate effects aren't the only concerns when using pesticides. Pesticide residues can remain active for days, weeks, or, in some cases, months. Propoxur, the active ingredient in products such as Baygon, used to kill cockroaches, flies, and other household and lawn pests — and one of the most toxic pesticides available for home use — is a neurotoxin that can temporarily disrupt the function of the nervous system.

Signal words warn you only of possible immediate effects the product can have if not used according to instructions. They don't warn you about possible effects over time. Yet many maintenance products are used regularly — weekly or even daily, for years or even decades. Chronic exposure to propoxur, for example, may cause cancer. Phenol is a suspected carcinogen. Methylene chloride has been cited by the Consumer Product Safety Commission as "one of the highest cancer risks ever calculated for a chemical in a consumer product."

Environmental Products

The alternative products listed below are not risk free, but their immediate and long-term dangers pale in comparison to those of conventional brands.

Cleaners and Polishes
Some low-toxicity products have been on supermarket shelves for years. Twenty Mule Team Borax, a disinfectant, water softener, and laundry brightener that also controls odors and mold, is made from a naturally occurring mineral. Bon Ami cleansing powder, which has been around for about a century, contains none of the chlorine in conventional scouring powders. Murphey's Oil Soap is made with only natural ingredients and fragrances.

A variety of "green" products are now available at health food stores and from mail-order suppliers. Some are derived from natural, renewable substances, from citrus solvent to beeswax. Others are petrochemical based, like most conventional products, but they're specially formulated to be less irritating, polluting, or toxic. Most of these products are sold in packaging made of recycled and/or recyclable material.

Laundry and Dishwashing Products
The combined effect on the environment of 250 million Americans (and billions more worldwide) washing their clothes each week is staggering. In recent decades, dishwashing detergents have added to the pollutant load on lakes, streams, and other surface waters.

Two types of products are available: soaps and detergents. During the Second World War, when the oils and fats used in making soap were scarce, petrochemical detergents were developed. Today, some detergents are vegetable based; however, even these are combined with a synthetic sulfuric acid molecule and the result is a material that does not exist in nature. Detergents can contain other synthetic and potentially toxic substances such as optical brighteners, solvents, and fragrances. Soap is less toxic than detergent. It biodegrades more quickly in sewage treatment plants and in the process produces fewer toxic by-products, making it less harmful to fish and other aquatic life.

Soap, however, has a big drawback: in hard water it reacts with minerals, leaving a grayish film on clothes. What's more, you need more soap to provide cleaning suds, as much of the soap becomes bound up with the minerals. Annie Berthold-Bond, author of *Clean and Green: The Complete Guide to Nontoxic and Environmentally Safe Housekeeping,* recommends using soap in conjunction with a natural water softener such as ½ cup baking soda, ¼ cup borax, ½ cup washing soda, or 2 tablespoons zeolite (a combination of minerals found in volcanic rock) for each full load.

Pest Controls

It is possible to control pests in and around the home without resorting to chemical warfare by using a series of common-sense measures collectively known as integrated pest management, or IPM. IPM admittedly is more labor intensive than simply reaching for the instant bug bomb. But it's a lot safer for you and your family.

IPM begins with a close look at the habitat and lifestyle of the pest, then proceeds with efforts to eliminate its habitat and disrupt its lifestyle. If chemical controls are needed, the least-poisonous pesticide is carefully chosen and judiciously used at the most vulnerable period in the pest's life cycle.

How do you go about practicing IPM?

- Identify the enemy. Sometimes, identification is easy. When in doubt, take a sample to your local Cooperative Extension or a pest-control company for a precise ID. This is critical in choosing which controls to use.

- Learn about the life cycle of the pest, its habits, where it feeds, sleeps, and hides. This helps determine what kind of control will be the most effective, and when. For example, if you sprinkle boric acid where roaches hide, they will walk through it, ingest it later when they "groom" themselves, and eventually die.

- Begin with nonchemical weapons. The first step in roach control, for example, is to try to eliminate the pest's habitat and food sources — by, say, plugging up holes in the wall and keeping the kitchen spotless. Next, try using non-toxic traps — roach "motels" and the like.

- The next escalation in tactics is using "biological controls." This sounds ominous but it simply means using one or more of a pest's natural predators to eradicate it for you. Nematodes, for example, tiny parasitic worms, have proven effective at eliminating subterranean termites, especially in areas with sandy soils.

- Use chemical pesticides, whether natural or synthetic, only as a last resort. A few plant-derived pesticides are now available, including rotenone, derived from the roots of many plants; ryania, made by grinding up the stems of a South American shrub; pyrethrum, which contains active ingredients extracted from the seeds of chrysanthemums; and sabadilla, which comes from the seeds of a South American lily. Natural pesticides are still poisons. For example, rotenone is more acutely toxic than most synthetic pesticides sold for use by homeowners. However, natural pesticides generally break down more quickly in the environment than synthetics and therefore do less ecological damage.

Entire volumes can be written on IPM. A number of good books are already available. Two excellent sources are *Common Sense Pest Control: Least-Toxic Solutions for Your Home, Garden, Pets and Community,* by William Olkowski, Sheila Daar, and Helga Olkowski (The Taunton Press, 1991) and the Brooklyn Botanic Garden's *Natural Pest Control* (1994).

Paint Strippers

Paint strippers can be highly toxic to breathe or to get on your skin — both of which you are quite likely to do when using them, even if you ventilate the area, wear gloves, and take other precautions. Until recently, most strippers used

WHAT TO DO WITH LEAD-BASED PAINT

Lead-based paint was widely used in home construction and decoration during the 1950s. Its use declined sharply during the 1970s and virtually ceased by 1980. However, the Agency for Toxic Substances and Disease Registry (ATSDR) estimates that there are about 42 million U.S. homes constructed before 1980 that have lead-based paint on interior or exterior surfaces. Both the increasing deterioration of this paint over time and the increase in renovations of older homes have made lead-based paint an urgent issue.

A report issued in 1988 by the ATSDR lists numerous health problems associated with elevated lead levels in the blood. Several studies have linked lead exposure to high blood pressure in middle-aged men. However, the groups most at risk are fetuses, infants, and children. Lead has an adverse effect on the brain and central nervous system of children; elevated levels can cause delayed cognitive development, reduced IQ scores, impaired hearing, and other problems (see Appendix A, "A Guide to Home Pollutants"). It was once believed that the major cause of elevated blood levels in children was the ingestion of chips of old lead paint. However, house dust contaminated with lead is now believed to be a bigger source of exposure.

If you suspect that old lead paint is in your house, have lead levels tested by a qualified professional using an X-ray fluorescence analyzer.

How do you get rid of old lead paint? Repainting over older coats of lead-based paint is *not* an effective way to deal with the problem because the lead can work its way through. The following removal methods have been prohibited in various federal and/or state regulations because they generate lead dust: gas-fired open-flame torches, sanding or dry scraping without an attached high-efficiency particulate absolute (HEPA) filtration vacuum apparatus, uncontained water blasting, and open abrasive blasting.

Chemical strippers don't generate dust and can be used in small areas, although they do raise other safety issues because the chemicals used are often toxic (see the list of less-toxic strippers on page 244). Encapsulation is the only measure recommended for large interior surfaces covered with old lead paint, such as walls, floors, and ceilings. What this means is you should cover these surfaces with wallboard, wood, or other material. If this is impractical, you should seek the guidance of reputable abatement professionals — don't undertake lead paint removal on such a large scale yourself. In any case, make sure children are out of the house and harm's way until all the work is done, and have dust levels tested before you move the kids back in.

by do-it-yourselfers were strong — and very toxic — volatile solvents like methylene chloride (see the section on volatile organic compounds in "A Guide to Home Pollutants," pages 245–249).

Nowadays there are safer strippers. Citristrip, for example, is made with a natural citrus solvent. Others, such as AFM's Lift Off, are synthetic formulations, but they're water-based instead of solvent-based, contain no methylene chloride, and are much less toxic than conventional products.

Do these safer strippers work? Yes, slowly but surely. They take longer to penetrate the layers of old paint. However, the greater safety of these products more than compensates for the slower action.

There are non-chemical ways to remove paints, too, but these have their own hazards. Using a torch or heat gun increases the risk of burns or fire. Sanding and scraping create dust and should never be used to remove old lead paint (see "What to Do with Lead-Based Paint," page 241).

Suppliers

CLEANERS AND POLISHES

AFM Enterprises, Inc.
1140 Stacy Court, Riverside, CA 92507; (714) 781-6860

Allen's Naturally
P.O. Box 514, Farmington, MI 48332; (800) 352-8971 or (313) 453-5410

Auro, Sinan Co.
P.O. Box 857, Davis, CA 95617-0857; (916) 753-3104

Cloverdale Inc.
P.O. Box 268, 5 Smith Place, West Cornwall, CT 06796; (800) 421-4818 or (203) 672-0216

The Dasun Company
P.O. Box 668, Escondido, CA 92033; (800) 433-8929 or (619) 480-8929

Earth Rite
23700 Mercantile Road, Beachwood, OH 44122; (800) 328-4408

Earth Wise
1790 30th Street, Boulder, CO 80301; (303) 447-0119

Eco Design
1365 Rufina Circle, Santa Fe, NM 87501; (800) 621-2591 or (505) 438-3448

Ecover
6-8 Knight Street, Norwalk, CT 06850; (203) 853-4166

Healthy Kleaner
P.O. Box 4656, Boulder, CO 80306; (800) EARTH-29

Life Tree
P.O. Box 1203, Sebastopol, CA 95472; (707) 577-0324

Natural Chemistry, Inc.
244 Elm Street, New Canaan, CT 06840; (800) 753-1233 or (203) 966-8761

Seventh Generation
49 Hercules Drive, Colchester, VT 05446-1672; (800) 456-1177

Pool and Spa Cleaners

Natural Chemistry, Inc., see above

LAUNDRY AND DISHWASHING PRODUCTS

Allen's Naturally, see above

Auro, Sinan Co., see above

Earth Rite, see above

Earth Wise, see above

Eco Design, see above

Ecover, see above

Life Tree, see above

Seventh Generation, see above

PEST-CONTROL PRODUCTS

The following companies offer a variety of natural and less-toxic pesticides, traps, beneficial insects, and other products:

Gardens Alive!
5100 Schenley Place, Lawrenceburg, IN 47025; (812) 537-8650

Harmony Farm Supply
P.O. Box 460, Graton, CA 95444; (707) 823-9125

Integrated Fertility Management
333 Ohme Gardens Road, Wenatchee, WA 98801; (800) 332-3179 or (509) 662-3179

Mellinger's, Inc.
2310 W. South Range Road, North Lima, OH 44452; (216) 549-9861

Natural Gardening Company
217 San Anselmo Avenue, San Anselmo, CA 94960; (415) 456-5060

Nature's Control
P.O. Box 35, Medford, OR 97501; (503) 899-8318

Necessary Organics
422 Salem Avenue, New Castle, VA 24127; (800) 447-5354 or (703) 864-5103

Pest Management Supply, Inc.
P.O. Box 938, Amherst, MA 01004; (800) 272-7672

Ringer Corporation
9959 Valley View Road, Eden Prairie, MN 55344; (800) 423-7544 or (612) 941-4180

Safer, Inc.
9959 Valley View Road, Eden Prairie, MN 55344; (800) 423-7544 or (612) 941-4180

The following companies offer less-toxic termite-control products and services:

Bio Integral Resource Center
P.O. Box 7414, Berkeley, CA 94707; (415) 524-2567
Publishes a booklet that includes articles on sand barriers, liquid nitrogen, and heat treatments, predatory nematodes, and other safe termite controls. Also publishes an annual list of product suppliers.

BioLogic Natural Pest Control
P.O. Box 177, Willow Hill, PA 17271; (717) 349-2789
Predatory nematodes.

N-Viro Products Ltd.
610 Walnut Avenue, Bohemia, NY 11716; (516) 567-2628
Predatory nematodes.

Tallon Termite and Pest Control
1949 East Market Street, Long Beach, CA
90805; (800) 779-2653 or (310) 422-1131
Predatory nematodes and liquid nitrogen.

The following companies sell boric acid, which
is effective against roaches:

Copper Brite, Inc.
P.O. Box 50610, Santa Barbara, CA 93150-
0610; (805) 565-1566

Organic Control, Inc.
5132 Venice Boulevard, Los Angeles, CA
90019; (213) 937-7444

R Value West
10926-B Grand Avenue, Temple City, CA
91780; (818) 448-4833

The following companies offer flea-control
products, including flea combs, diatomaceous
earth, pyrethrum powder, and flea shampoos
and sprays:

Natural Animal Inc.
P.O. Box 1177, St. Augustine, FL 32085; (800)
274-7387 or (904) 824-5884

Ringer Corporation, see above

Safer, Inc., see above

Sandoz Agro Inc.
1300 East Touhy Avenue, Des Plaines, IL
60018; (800) 527-0512 or (708) 699-1616

The following company sells copper mesh barriers for rodent control:

Allen Special Products, Inc.
P.O. Box 605, Montgomeryville, PA 18936;
(800) 848-6805 or (215) 997-9077

PAINT STRIPPERS
AFM Enterprises, Inc., see above

Auro, Sinan Co., see above

Eco Design, see above

**Citristrip, Specialty Environmental Technolo-
gies, Inc.**
4520 Glenmeade Lane, Auburn Hills, MI
48326; (800) 899-0401 or (810) 340-0400

Lift Off, AFM Enterprises, Inc., see above

EXTERIOR CLEANING PRODUCTS
Armor All
4055 Faber Place Drive, Charleston, SC
29405; (800) 398-3892 or (803) 566-0766
E-ZD Prepaint House Wash, E-ZD Vinyl Siding
Wash, and E-ZD Deck Wash.

A GUIDE TO HOME POLLUTANTS

A growing body of scientific evidence suggests that the air inside our homes and workplaces is often far more polluted than the air outside, even in the largest and most industrialized cities. Because most people spend about 85 percent of their time inside, indoor air pollution may be a much greater health risk than air pollution outdoors. What's more, those typically exposed to indoor air pollutants for the longest periods of time are often the most vulnerable to their adverse effects — children, the elderly, and the chronically ill.

The levels of pollutants emitted by individual sources may not pose a significant health threat. However, most homes have many sources of indoor pollution. The cumulative effects of these pollutants can pose serious health risks.

Although it has been regulating outdoor air quality for about twenty years, the U.S. Environmental Protection Agency does not regulate indoor air pollution, although it has come up with recommended maximum levels of some pollutants, such as radon. But by choosing home building, decorating, and maintenance products carefully and making sure your house is adequately ventilated, you can dramatically improve the air quality in your home and reduce the health risks for you and your family. The following is a concise summary of the major indoor pollutants, their health effects, and the materials most likely to emit them.

ASBESTOS

Asbestos is a naturally occurring mineral that was once used widely as an insulator and fireproofing material.

Sources: Deteriorating or damaged insulation or fireproofing materials.

Health effects: No immediate symptoms. Inhalation of asbestos has been shown to result in lung disease, including fibrosis and scarring of the lower lobes of the lungs, lung cancers, and abdominal cancers.

BIOLOGICAL CONTAMINANTS

These include fungi, molds and mildews, pollen, animal dander, and dust mites.

Sources: Wet or moist walls, ceilings, carpets, and furniture; poorly maintained humidifiers, dehumidifiers, and air conditioners; bedding; pets. These contaminants tend to collect on wall-to-wall carpeting, fabrics, and upholstered furniture.

Health effects: Eye, nose, and throat irritation; shortness of breath; dizziness; fatigue; fever; asthma.

COMBUSTION GASES

Complete combustion of fuels used for heating and cooking in the home produces water and carbon

Toxin — A toxin is a poison, a substance with the capacity to harm. A substance's toxicity is usually rated according to the dose that is required to kill 50 percent of a test group of animals; this rating is called the LD50 (LD stands for "lethal dose"). However, a toxin's effects may not be lethal. Sublethal effects often damage specific organs such as the liver or brain.

Carcinogen — A carcinogen is a substance that causes cancer. The effects of substances that are not carcinogenic will occur only above a given threshold of exposure. However, there is no known level below which carcinogens are harmless.

Health effects from indoor air pollutants fall into two categories:

Acute — A reaction that occurs within twenty-four hours after exposure.

Chronic — A reaction that takes three months or more to manifest itself.

Route of exposure — The amount of harm that a toxin can do depends on the route of exposure. Exposure can occur through contact with the skin or eyes, through inhalation, or through ingestion. Some toxins commonly found in indoor air can enter the body through contact with the skin or eyes and through ingestion, as well as through inhalation. For example, petrochemical solvents can be inhaled, can pass through the skin into the bloodstream, and can be ingested in drinking water. All three types of exposure are potentially hazardous.

---

dioxide. However, no appliance is totally efficient. Incomplete burning of fuels produces many combustion products, including irritant gases such as carbon monoxide and nitrogen dioxide.

Sources: Unvented kerosene heaters, gas stoves, and gas heaters; leaking chimneys and furnaces; downdrafting from wood stoves and fireplaces. Also, tobacco smoke and automobile exhaust from attached garages.

Health effects: At low levels, carbon monoxide causes fatigue in healthy people and chest pain in those with heart disease. At higher levels, symptoms include impaired vision and coordination, headaches, dizziness, confusion, nausea, and flu-like symptoms. At very high levels it can be fatal. Fortunately, inexpensive home carbon monoxide detectors, which will warn you when levels of this potentially lethal gas are high, are available at hardware stores and home supply centers. Nitrogen dioxide causes eye, nose, and throat irritation. It can cause impaired lung function and increase young children's susceptibility to respiratory infection.

ELECTROMAGNETIC FIELDS

An electric field occurs whenever charges are present; a magnetic field occurs whenever charges are in motion, as is the case of an electric current. Because the two types of field are often present together, they are referred to jointly as electromagnetic fields. These fields, in varying magnitudes and at different times, are found in virtually every home and office. The most common unit of measure for a magnetic field is the milliGauss, or mG.

Sources: EMFs are emitted by all wires and machines that carry current, including home wiring and all electric-powered home appliances, from coffeemakers to VCRs, as well as power lines and electrical substations. U.N. World Health Organization measurements of electromagnetic fields very close to various appliances ranged from a low of 25 to 500 mG for televisions to a high of 20,000 mG for hair dryers and can openers; levels fall off dramatically at a distance of about three feet. EMFs from powerlines, however, can remain strong to a distance of 100 feet or more.

Health effects: Scientists have yet to agree on the hazards posed by electromagnetic fields. A number of respected scientists contend that EMFs are linked to the development of cancer, particularly in children, while at least an equal number of equally respected scientists contend that no link has been established. Two recent Swedish studies, which show the strongest connection yet between EMFs and leukemia in adults and children, have prompted the Swedish government to begin work on regulations to limit human exposure.

ENVIRONMENTAL TOBACCO SMOKE

Environmental tobacco smoke is the fancy term for secondhand smoke that is inhaled by nonsmokers who share living or work spaces with a smoker or smokers.

Sources: Cigarette, pipe, and cigar smoking.

Health effects: Eye, nose, and throat irritation; headaches; bronchitis; pneumonia. Increases the risk of respiratory and ear infections in children. A growing body of evidence also suggests that secondhand tobacco smoke causes cancer.

FORMALDEHYDE

Formaldehyde is one of the most ubiquitous volatile organic compounds found indoors (see "Volatile Organic Compounds," below). It is used widely in the manufacture of building materials and household products. It is also a by-product of combustion.

Sources: The scores of sources of formaldehyde in the typical home include hardwood plywood wall paneling; particleboard and fiberboard, including furniture made with them; permanent press fabrics, bedding, and clothing; glues and adhesives; paints; and urea-formaldehyde foam insulation (although it was widely installed during the 1970s, few if any homes are now being insulated with this product). Most studies indicate that formaldehyde emissions will usually decrease as products age.

Health effects: Eye, nose, and throat irritation; wheezing and coughing; fatigue; skin rash; severe allergic reactions. Formaldehyde has been shown to cause cancer in animals and may cause cancer in humans. There is also evidence that some people can develop chemical sensitivity after exposure to formaldehyde.

LEAD

Scientists have long known that lead is a harmful environmental pollutant. Fetuses, infants, and children are especially vulnerable to lead exposure.

Sources: Lead-based paint has been recognized for decades as a hazard to children who eat paint chips containing lead. A 1988 National Institute of Building Sciences task force report found that harmful exposures to lead can be created when lead-based paint is removed by sanding or open-flame burning; the resulting lead particles adhere to dust and can be inhaled. Drinking water can be another significant source of lead exposure. Lead leaches into the water as it passes through the old lead pipes or nonlead pipes connected with lead solder found in many older homes.

Health effects: Impaired mental and physical development in both fetuses and young children. Decreased coordination and mental abilities. Damage to kidneys, nervous system, and red blood cells. High blood pressure.

PESTICIDES

When you think about pesticides, what probably comes to mind are the chemicals used to eradicate weeds and pests outdoors in the lawn and garden. But a variety of poisons are routinely used indoors to kill household pests. Also, products used on lawns and gardens can drift or be tracked into the house. In fact, research shows widespread presence of pesticide residues in American homes.

Sources: Sprays and powders used to control fleas, roaches, termites, and other household pests; pest strips; mildewcides.

Health effects: Irritation of eyes, nose, and throat. Damage to central nervous system and kidneys. In some cases, cancer.

RADON

Radon is a naturally occurring, odorless, tasteless radioactive gas. It is produced from the radioactive decay of radium, which is formed from the decay of uranium. Since radium and uranium are ubiquitous elements in rock and soil, radon is, too. However, radon is present at excessively high-risk levels in some geological formations and therefore some areas of the country. Because radon is inert and thus not chemically bound or attached to other materials, it can move freely through very small spaces, such as those between particles of soil and rock.

Sources: Radon migrates from soil and rock through cracks in the basement or slab and into the indoor air. It can also migrate into groundwater and be released into household air when the tap is on or the shower is running.

Health effects: No immediate symptoms. However, the Environmental Protection Agency and the Surgeon General estimate that exposure to radon in homes and buildings causes between five thousand and twenty thousand cases of lung cancer each year. Smokers are at higher risk of developing radon-induced lung cancer.

RESPIRABLE PARTICLES

Respirable particles are generally defined as small particles (less than 10 microns in diameter) that are easily inhaled and can lodge deep in the lungs. They are produced by all combustion processes.

Sources: Fireplaces; wood-burning stoves; kerosene heaters; tobacco smoke.

Health effects: Eye, nose, and throat irritation; respiratory infections and bronchitis; lung cancer.

VOLATILE ORGANIC COMPOUNDS

The term volatile organic compounds refers collectively to a large number of mostly petrochemical-derived substances that readily volatilize, or become a breathable gas, at room temperatures and therefore contaminate indoor air. Hundreds of VOCs have been identified in household air, including formaldehyde, benzenes, toluene, styrene, xylenes, and chlorinated solvents such as trichloroethylene, carbon tetrachloride, methylene chloride, and chloroform.

Sources: Hundreds of household products, including paints, paint strippers, sealers and finishes, solvents, adhesives, cleansers and disinfectants, aerosol sprays, air fresheners, moth repellants, and dry-cleaned clothing. Certain materials, however, are more significant sources of VOCs than others. These include carpeting, vinyl flooring, particleboard, insulation, and "wet-applied" products such as adhesives, paints and stains, caulks, sealers and finishes, pesticides, and joint compound. Wet-applied products are particularly worrisome because such a large fraction of their content must evaporate into the air. Although most studies suggest that VOC emissions decrease over time, one study of joint compound conducted by Harvard University for the Environmental Protection Agency indicates that emissions can *increase* over time under certain temperature and humidity conditions.

Health effects: The health effects of volatile organic compounds from building and decorating materials are not well understood, especially the complex mixtures of VOCs found in indoor air. However, many individual VOCs are known or suspected human irritants and carcinogens. In one study, Lars Molhave of Denmark found that 82 percent of commonly emitted VOCs are known or suspected mucous membrane or eye irritants, and 25 percent are known or suspected human cancer causers. Short-term health effects can include eye, nose, and throat irritation, headaches, and loss of coordination. Long-term effects can include damage to the liver, kidney, and nervous system, as well as cancer.

FURTHER READING

The following publications and organizations, listed alphabetically, are some of the best sources of reliable, up-to-the-minute information on environmental home design and products:

American Council for an Energy-Efficient Economy
1001 Connecticut Avenue NW, Suite 801
Washington, D.C. 20036
(202) 429-8873
Publishes *The Most Energy Efficient Appliances* and *The Consumer Guide to Home Energy Savings.*

Bio-Integral Resource Center (BIRC)
P.O. Box 7414
Berkeley, CA 94707
(415) 524-2567
Publishes two newsletters and numerous other publications on safe pest control.

Brooklyn Botanic Garden
1000 Washington Avenue
Brooklyn, NY 11225
(718) 622-4433, ext. 274
Publishes numerous affordable books on natural gardening for readers across the United States and Canada, including *The Environmental Gardener, Natural Insect Control: The Ecological Gardener's Guide to Foiling Pests,* and *Going Native: Biodiversity in Our Own Backyards.* For information on how to order, call the number listed above. For information on how to subscribe to their 21st-Century Gardening Series of handbooks and their newsletter,

Plants & Gardens News, call (718) 622-4433, ext. 261.

Center for Resourceful Building Technology
P.O. Box 100
Missoula, MT 59806
(406) 549-7678
Publishes the *Guide to Resource-Efficient Building Elements,* with an emphasis on products made from recycled materials.

Clean & Green
by Annie Berthold-Bond
published by Ceres Press
P.O. Box 87
Woodstock, NY 12498
(914) 679-5573
One of the best guides to less-toxic home maintenance.

Environmental Building News
RR1, Box 161
Brattleboro, VT 05301
(802) 257-7300
An excellent source of timely information on sustainable design and construction.

Indoor Air Bulletin
2548 Empire Grade
Santa Cruz, CA 95060
(408) 425-3947
The place to turn for level-headed, non-alarmist information on indoor air pollution.

Interior Concerns
P.O. Box 2386
Mill Valley, CA 94942
(415) 389-8049
A newsletter and resource guide on interior design and products.

Microwave News
P.O. Box 1799, Grand Central Station
New York, NY 10163
(212) 517-2800
The best source of information on electromagnetic radiation. Publishes an updated list of manufacturers of gaussmeters, which you can use to measure electromagnetic fields in your home.

The Naturally Elegant Home
by Janet Marinelli with Robert Kourik
published by Little, Brown and Company
(800) 759-0190 to order
Features gorgeous environmental houses and gardens from coast to coast. Also includes a multitude of practical tips on making your own home both environmentally sound and beautiful.

Rocky Mountain Institute
1739 Snowmass Creek Road
Snowmass, CO 81654-9199
(303) 927-3851

Publishes *The Efficient House Sourcebook* and numerous other publications on energy- and water-efficient home technologies.

The Smart Kitchen: How to Design a Comfortable, Safe, Energy-Efficient and Environment-Friendly Workspace
by David Goldbeck
published by Ceres Press
P.O. Box 87
Woodstock, NY 12498
(914) 679-5573
An excellent primer on designing a "green" kitchen.

Woodworkers Alliance for Rainforest Protection (WARP)
289 College Street
Burlington, VT 05401
(802) 862-4448
Publishes the newsletter *Understory,* one of the finest sources of information on sustainably produced woods, both domestic and tropical.

INDEX